Huxley Thomas Henry, Martin Henry Newell

A Course of Practical Instruction in Elementary Biology

Huxley Thomas Henry, Martin Henry Newell

A Course of Practical Instruction in Elementary Biology

ISBN/EAN: 9783337216863

Printed in Europe, USA, Canada, Australia, Japan

Cover: Foto ©berggeist007 / pixelio.de

More available books at **www.hansebooks.com**

Cambridge:
PRINTED BY C. J. CLAY, M.A.
AT THE UNIVERSITY PRESS.

175975

PREFACE.

VERY soon after I began to teach Natural History, or what we now call Biology, at the Royal School of Mines, some twenty years ago, I arrived at the conviction that the study of living bodies is really one discipline, which is divided into Zoology and Botany simply as a matter of convenience; and that the scientific Zoologist should no more be ignorant of the fundamental phenomena of vegetable life, than the scientific Botanist of those of animal existence.

Moreover, it was obvious that the road to a sound and thorough knowledge of Zoology and Botany lay through Morphology and Physiology; and that, as in the case of all other physical sciences, so in these, sound and thorough knowledge was only to be obtained by practical work in the laboratory.

M.   *b*

The thing to be done, therefore, was to organize a course of practical instruction in Elementary Biology, as a first step towards the special work of the Zoologist and Botanist. But this was forbidden, so far as I was concerned, by the limitations of space in the building in Jermyn Street, which possessed no room applicable to the purpose of a laboratory; and I was obliged to content myself, for many years, with what seemed the next best thing, namely, as full an exposition as I could give of the characters of certain plants and animals, selected as types of vegetable and animal organization, by way of introduction to systematic Zoology and Palæontology.

In 1870, my friend Professor Rolleston, of Oxford, published his "*Forms of Animal Life.*" It appears to me that this exact and thorough book, in conjunction with the splendid appliances of the University Museum, leaves the Oxford student of the fundamental facts of Zoology little to desire. But the Linacre Professor wrote for the student of Animal life only, and, naturally, with an especial eye to the conditions which obtain in his own University; so that there was still room left for a Manual of wider scope, for the use of learners less happily situated.

In 1872 I was, for the first time, enabled to carry my own notions on this subject into practice, in the excellent rooms provided for biological instruction in the New Buildings at South Kensington. In the short course of Lectures given to Science Teachers on this occasion, I had the great advantage of being aided by my friends Dr Foster, F.R.S., Prof. Rutherford, F.R.S., and Prof. Lankester, F.R.S., whose assistance in getting the laboratory work into practical shape was invaluable.

Since that time, the biological teaching of the Royal School of Mines having been transferred to South Kensington, I have been enabled to model my ordinary course of instruction upon substantially the same plan.

The object of the present book is to serve as a laboratory guide to those who are inclined to follow upon the same road. A number of common and readily obtainable plants and animals have been selected in such a manner as to exemplify the leading modifications of structure which are met with in the vegetable and animal worlds. A brief description of each is given; and the description is followed by such detailed instructions as, it is hoped, will enable the student to know, of his own knowledge, the chief facts mentioned in the account of the animal or plant. The terms used in Biology will thus be represented by clear and definite images of the things to which they apply; a comprehensive, and yet not vague, conception of the phenomena of Life will be obtained; and a firm foundation upon which to build up special knowledge will be laid.

The chief labour in drawing up these instructions has fallen upon Dr Martin. For the general plan used, and the descriptions of the several plants and animals, I am responsible; but I am indebted for many valuable suggestions and criticisms from the botanical side to my friend Prof. Thiselton Dyer.

T. H. H.

LONDON,
September, 1875.

CONTENTS.

V.

MOULDS.

VI.

STONEWORTS.

VII.

THE BRACKEN FERN.

VIII.

THE BEAN PLANT.

IX.

THE BELL ANIMALCULE.

I.

YEAST (*Torula* or *Saccharomyces Cerevisiæ*).

YEAST is a substance which has been long known on account of the power which it possesses of exciting the process termed *fermentation* in substances which contain sugar.

If strained through a coarse filter, it appears to the naked eye as a brownish fluid in which no solid particles can be discerned. When some of this fluid is added to a solution of sugar and kept warm, the mixture soon begins to disengage bubbles of gas and become frothy; its sweetness gradually disappears; it acquires a spirituous flavour and intoxicating qualities; and it yields by distillation a light fluid—*alcohol* (or spirits of wine) which readily burns.

When dried slowly and at a low temperature, yeast is reduced to a powdery mass, which retains its power of exciting fermentation in a saccharine fluid for a considerable period. If yeast is heated to the temperature of boiling water, before it is added to the saccharine fluid, no fermentation takes place; and fermentation which has commenced is stopped by boiling the saccharine liquid.

A saccharine solution will not ferment spontaneously. If it begins to ferment, yeast has undoubtedly got into it in some way or other.

If the yeast is not added directly to the saccharine fluid, but is separated from it by a very fine filter, such as porous earthenware, the saccharine fluid will not ferment, although the filter allows the fluid part of the yeast to pass through into the solution of sugar.

M.   1

If the saccharine fluid is boiled, so as to destroy the efficiency of any yeast it may accidentally contain, and then allowed to come in contact only with such air as has been passed through cotton wool, it will never ferment. But if it is exposed freely to the air, it is almost sure to ferment sooner or later, and the probability of its so doing is greatly increased if there is yeast anywhere in the vicinity.

These experiments afford evidence (1) that there is something in yeast which provokes fermentation, (2) that this something may have its efficiency destroyed by a high temperature, (3) that this something consists of particles which may be separated from the fluid which contains them by a fine filter, (4) that these particles may be contained in the air; and that they may be strained off from the air by causing it to pass through cotton wool.

Microscopic examination of a drop of yeast shews what the particles in question are.

Even with a hand-glass, the drop no longer appears homogeneous, as it does to the naked eye, but looks as if fine grains of sand were scattered through it; but a considerable magnifying power (5—600 diameters) is necessary to shew the form and structure of the little granules which are thus made visible. Under this power, each granule (which is termed a *Torula*) is seen to be a round, or oval, transparent body, varying in diameter from $\frac{1}{2800}$th to $\frac{1}{7000}$th of an inch (on the average about $\frac{1}{3000}$th).

The *Torulæ* are either single, or associated in heaps or strings. Each consists of a thin-walled sac, or bag, containing a semi-fluid matter, in the centre of which there is often a space full of a more clear and watery fluid than the rest, which is termed a 'vacuole.' The sac is comparatively tough, but it may be easily burst, when it gives exit to its contents, which readily diffuse themselves through the surrounding fluid. The whole structure is called a 'cell;' the sac being the 'cell-wall' and the contents the 'protoplasm.'

When yeast is dried and burned in the open air it gives rise to the same kind of smell as burning animal matter, and a certain quantity of mineral ash is left behind. Analysed into its chemical elements, yeast is found to contain Carbon, Hydrogen, Oxygen, Nitrogen, Sulphur, Phosphorus, Potassium, Magnesium and Calcium; the last four in very small quantities.

These elements are combined in different ways, so as to form the chief proximate constituents of the *Torula*, which are (1) a Protein compound, analogous to Casein, (2) Cellulose, (3) Fat, and (4) Water. The cell-wall contains all the Cellulose and a small proportion of the mineral matters. The protoplasm contains the Protein compound and the Fat with the larger proportion of the mineral salts.

These *Torulæ* are the 'particles' in the yeast which have the power of provoking fermentation in sugar; it is they which are filtered off from the yeast when it loses its efficiency by being strained through porous earthenware; it is they which form the fine powder to which yeast is reduced by drying, and which, from their extreme minuteness, are readily diffused through the air in the form of invisible dust.

That the *Torulæ* are living bodies is proved by the manner in which they grow and multiply. If a small quantity of yeast is added to a large quantity of clear saccharine fluid so as hardly to disturb its transparency, and the whole is kept in a warm place, it will gradually become more and more turbid, and, after a time, a scum of yeast will collect, which may be many thousand, or million, times greater in weight than that which was originally added. If the *Torulæ* are examined as this process of multiplication is going on, it

will be found that they are giving rise to minute buds, which
rapidly grow, assume the size of the parent *Torula,* and
eventually become detached; though, generally, not until
they have developed other buds, and these yet others. The
Torulæ thus produced by gemmation, one from the other,
are apt long to adhere together, and thus the heaps and
strings mentioned, as ordinarily occurring in yeast, are produced. No *Torula* arises except as the progeny of another;
but, under certain circumstances, multiplication may take
place in another way. The *Torula* does not throw out a bud,
but its protoplasm divides into (usually) four masses, termed
ascospores, each of which surrounds itself with a cell-wall,
and the whole are set free by the dissolution of the cell-
wall of the parent. This is multiplication by *endogenous
division.*

As each of the many millions of *Torulæ* which may thus
be produced from one *Torula* has the same composition as
the original progenitor, it follows that a quantity of Protein,
Cellulose and Fat proportional to the number of *Torulæ*
thus generated, must have been produced in the course of
the operation. Now these products have been manufactured
by the *Torulæ* out of the substances contained in the fluid
in which they float and which constitute their food.

To prove this it is necessary that this fluid should have
a definite composition. Several fluids will answer the purpose, but one of the simplest (Pasteur's solution) is the
following.

Water.................. (H_2O).
Sugar .... ($C_{12}H_{22}O_{11}$).
Ammonium Tartrate ($C_4H_4(NH_4)_2O_6$).
Potassium Phosphate (KH_2PO_4).
Calcium Phosphate ($Ca_3P_2O_8$).
Magnesium Sulphate ($MgSO_4$).

In this fluid the *Torula* will grow and multiply. But it will be observed that the fluid contains neither Protein nor Cellulose, nor Fat, though it does contain the elements of these bodies arranged in a different manner. It follows that the *Torula* must absorb the various substances contained in the water and arrange their elements anew, building them up into the complex molecules of its own body. This is a property peculiar to living things.

The *Torula* being alive, the question arises whether it is an animal or a plant. Although no sharp line of demarcation can be drawn between the lowest form of animal and of vegetable life, yet *Torula* is an indubitable plant, for two reasons. In the first place, its protoplasm is invested by a continuous cellulose coat, and thus has the distinctive character of a vegetable cell. Secondly, it possesses the power of constructing Protein out of such a compound as Ammonium Tartrate, and this power of manufacturing Protein is distinctively a vegetable peculiarity. *Torula* then is a plant, but it contains neither starch nor chlorophyll, it absorbs oxygen and gives off carbonic anhydride, thus differing widely from the green plants. On the other hand, it is, in these respects, at one with the great group of *Fungi*. Like many of the latter, its life is wholly independent of light, and in this respect, again, it differs from the green plants.

Whether *Torula* is connected with any other form of *Fungi* is a question which must be left open for the present. It is sufficient to mention the fact that under certain circumstances some Fungi (e.g. *Mucor*) may give rise to a kind of *Torula* different from common yeast.

The fermentation of the sugar is in some way connected with the living condition of the *Torula*, and is arrested by all those conditions which destroy the life of the *Torula* and prevent its growth and reproduction. The greater part of the sugar is resolved into Carbonic anhydride and Alcohol, the elements of which, taken together, equal in weight those of the sugar. A small part breaks up into Glycerine and Succinic acid, and one or two per cent. is not yet accounted for, but is perhaps assimilated by the *Torula.*

This is the more probable as *Torula* will grow and multiply actively in a solution in which sugar and Ammonium Nitrate replace the Ammonium Tartrate of the former solution, in which case the carbon of the Protein, Cellulose and Fat manufactured, must be obtained from the sugar. Moreover, though oxygen is essential to the life of the *Torula*, it can live in saccharine solutions which contain no free oxygen, appearing, under these circumstances, to get its oxygen from the sugar.

It has further been ascertained that *Torulæ* flourish remarkably in solutions in which sugar and pepsin replace the Ammonium Tartrate. In this case, the nitrogen of their protein compounds must be derived from the pepsin; and it would seem that the mode of nutrition of such *Torulæ* approaches that of animals.

LABORATORY WORK.

Sow some fresh baker's yeast in Pasteur's fluid[1] with

[1] Pasteur's fluid:

Potassium Phosph.	20	parts.
Calcium Phosph.	2	,,
Magnesium Sulphate	2	,,
Ammonium Tartrate	100	,,
[Cane Sugar	1500	,,]
Water	8576	,,
	10,000	parts.

The sugar is to be omitted when Pasteur's fluid "without sugar" is ordered. Pasteur himself used actual yeast ash; the above constituents give an imitation ash, which, with the ammonium salt and sugar, answers all practical purposes.

sugar and keep it in a warm place : as soon as the mixture begins to froth up, and the yeast is manifestly increasing in quantity, it is ready for examination.

A. MORPHOLOGY.

1. Spread a little out, on a slide, in a drop of the fluid, and examine it with a low power ($\frac{1}{2}$ inch objective, Hartnack, No. 4) *without a cover-glass.* Note the varying size of the cells, and their union into groups.

2. Cover a similar specimen with a thin glass and examine it under a high power ($\frac{1}{8}$ objective. Hartnack, No. 7 or 8, Oc. 3 or 4).

    a. Note the size (measure), shape, surface and mode of union of the cells.

    b. Their structure : sac, protoplasm, vacuole.

        α. *Sac;* homogeneous, transparent.

        β. *Protoplasm;* less transparent ; often with a few clear shining dots in it.

        γ. *Vacuole;* sometimes absent ; size, position.

        δ. The relative proportion of sac, protoplasm, and vacuole in various cells.

    Draw a few cells carefully to scale.

3. Run in magenta solution under the cover-glass. (This is readily done by placing a drop of magenta solution in contact with one side of the cover-glass, and a small strip of blotting paper at the opposite side.)

    a. Note what cells stain soonest and most deeply, and what part of each cell it is that stains : the sac is unaffected ; the protoplasm stained ; the vacuole unstained, though it frequently appears pinkish, being seen through a coloured layer of protoplasm.

4. Burst the stained cells by placing a few folds of blotting paper on the surface of the cover-glass and pressing smartly with the handle of a mounted needle: note the torn empty and colourless, but solid and uncrushed transparent sacs; the soft crushed stained protoplasm.

5. Repeat observation 3, running in iodine solution instead of magenta. The protoplasm stains brown; the rest of the cell remains unstained. Note the absence of any blue coloration; *starch is therefore not present.*

6. Treat another specimen with potash solution, running it in as before: this reagent dissolves out the protoplasm, leaving the sac unaltered.

7. [Sow a few yeast-cells in Pasteur's solution in a moist chamber and keep them under observation from day to day; watch their growth and multiplication.]

8. [Endogenous division: take some dry German yeast; suspend it in water and shake so as to wash it. Let the mixture stand for half an hour: pour off the supernatant fluid, and, with a camel's hair pencil, spread out the creamy deposit in a thin layer on fresh cut potato slices or on a plate of plaster of Paris, and place with wet blotting paper under a bell-jar: examine from day to day with a very high power (800 diam.) for *ascospores*, which will probably be found on the eighth or ninth day.]

B. PHYSIOLOGY.

(Conditions and results of the vital activity of Torula.)

1. Sow a fair-sized drop of yeast in—

a. Distilled water.

b. 10 per cent. solution of sugar in water.

c. Pasteur's fluid without the sugar.

    d. Pasteur's fluid with sugar.

    [*e.* Mayer's pepsin solution[1].]

Keep all at about 35° C., and compare the growth of the yeast, as measured by the increase of the turbidity of the fluid, in each case. "*a*" will hardly grow at all, "*b*" better, "*c*" better still, "*d*" well, and "*e*" best of all. Note that bubbles of gas are plentifully evolved from both the solutions which contain sugar.

That any growth at all takes place, in the case of experiments *a* and *b*, is due to the fact that the drop of yeast added contains nutritious material sufficient to provide for that amount of growth.

2. Prepare two more specimens of "*d*" and keep one in a cold—the other in a warm (35° C.) place, but otherwise under like conditions. Compare the growth of the yeast in the two cases; it is much greater in the specimen kept warm.

3. Prepare two more specimens of "*d*"; keep both warm, but one in darkness, the other exposed to the light: that in the dark will grow as well as the other; sunlight is therefore not essential to the growth of Torula.

4. Sow some yeast-cells in Pasteur's solution in a flask, the neck of which is closed by a plug of cotton wool, and boil for five minutes; then set it aside; no signs of vitality will afterwards be manifested by the yeast in the flask; it is *killed* by exposure to this temperature.

---

[1] Mayer's solution (with pepsin) =

15 per cent. solution of sugar-candy	20 cc.
Dihydropotassic phosphate	0· 1 grm.
Calcic phosphate	0· 1 grm.
Magnesic sulphate	0· 1 grm.
Pepsin	0·23 grm.

5. [Take two test tubes; in one place some yeast, with Pasteur's solution containing sugar; in the other place baryta water, and then connect the two test tubes by tightly fitting perforated corks and a bent tube passing from above the surface of the fluid in the first tube to the bottom of the baryta water in the second; pass a narrow bent tube, open at both ends, through the cork of the baryta water tube, so that its outer end dips just below the surface of some solution of potash[1]. All gas formed in the first tube will now bubble through the baryta water in the second, and, from thence, any that is not absorbed will pass out through the potash into the air. An abundant precipitate of barytic carbonate will be formed which can be collected and tested. The fermenting fluid, therefore, evolves carbonic anhydride.]

6. [Grow some yeast in Pasteur's solution (with sugar), in a nearly closed vessel (say a bottle with a cork through which a long narrow open tube passes): as soon as the evolution of gas seems to have ceased, distil the fluid in a water bath and condense and collect the first fifth that comes over: redistil this after saturation with potassic carbonate, and test the distillate for alcohol by its odour and inflammability, and by the sulphuric acid and potassic dichromate test.]

7. [Determine that heat is evolved by a fluid in which active alcoholic fermentation is going on. Place 200 cc. of fresh yeast in a flask, and add 1 litre of Pasteur's fluid with sugar: put another litre of the fluid alone in a similar flask, cover each flask with a cloth and place the two side by side in a place protected from draughts. When gas begins to be actively evolved from the yeast-containing solution, take the temperature of the fluid in each flask with a good thermometer; the temperature of the one in which fermentation is going on will be found the higher.]

---

[1] The object of the potash is to shield the baryta water from any carbonic anhydride that may be in the atmosphere.

PROTOCOCCUS (*Protococcus pluvialis*).

IF the mud which accumulates in roof-gutters, water-butts, and shallow pools, be collected, it will be found to contain, among many other organisms, specimens of *Protococcus*. In one of the two conditions in which it occurs, *Protococcus* is a spheroidal body $\frac{1}{350}$ to $\frac{1}{10000}$ of an inch in diameter, composed, like *Torula*, of a structureless tough transparent wall, inclosing viscid and granular protoplasm. The chief solid constituent of the cell-wall is cellulose. The protoplasm contains a nitrogenous substance, doubtless of a proteinaceous nature, though its exact composition has not been determined, and indications of starchy matter are sometimes to be found in it. Either diffused through it, or collected in granules, is a red or green colouring matter (*Chlorophyll*). Individual *Protococci* may be either green or red; or half green and half red; or the red and green colours may coexist in any other proportion.

In addition to the single cells, others are found divided by partitions, continuous with the cellulose wall, into two or more portions, and the cells thus produced by *fission* become separate, and grow to the size of that form from which they started. In this manner *Protococcus* multiplies with very great rapidity. Multiplication by gemmation in the mode observed in *Torula* is said to occur, but is certainly of rare occurrence.

The influence of sunlight is an essential condition of the growth and multiplication of *Protococcus;* under that influence, it decomposes carbonic anhydride, appropriates the carbon, and sets oxygen free. It is this power of obtaining the carbon which it needs from carbonic anhydride, which is the most important distinction of *Protococcus*, as of all plants which contain chlorophyll, from *Torula* and the other *Fungi.*

As *Protococcus* flourishes in rain-water, and rain-water contains nothing but carbonic anhydride, which it absorbs along with other constituents of the atmosphere, ammonium salts (usually ammonium nitrate, also derived from the air) and minute portions of earthy salts which drift into it as dust, it follows that it must possess the power of constructing protein by rearrangement of the elements supplied to it by their compounds. *Torula*, on the other hand, is unable to construct protein matter out of such materials as these.

Another difference between *Torula* and *Protococcus* is only apparent: *Torula* absorbs oxygen and gives out carbonic anhydride; while *Protococcus*, on the contrary, absorbs carbonic anhydride and gives out oxygen. But this is true only so long as the *Protococcus* is exposed to sunlight. In the dark, *Protococcus*, like all other living things, undergoes oxidation and gives off carbonic anhydride; and there is every reason to believe that the same process of oxidation and evolution of carbonic anhydride goes on in the light, but that the loss of oxygen is far more than covered by the quantity set free by the carbon-fixing apparatus, which is in some way related to the chlorophyll.

The still condition of *Protococcus*, just described, is not the only state in which it exists. Under certain circumstances, a *Protococcus* becomes actively locomotive. The protoplasm withdraws itself from the cell-wall at all but two points, where it protrudes through the wall in the form of long vibratile filaments or *cilia*, and by the lashing of these cilia the cell is propelled with a rolling motion through the water. The movement of the cilia is so rapid, and their substance is so transparent and delicate, that they are invisible until they begin to move slowly, or are treated with reagents, such as iodine, which colour them.

Not unfrequently the cell-wall eventually vanishes, and the naked protoplasm of the cell swims about, and may undergo division and multiplication in this state. Sooner or later, the locomotive form draws in its cilia, becomes globular, and, throwing out a cellulose coat, returns to the resting state.

For reasons similar to those which prove the vegetable nature of *Torula*, *Protococcus* is a plant, although, in its locomotive condition, it is curiously similar to the Monads among the lowest forms of animal life. But it is now known that many of the lower plants, especially in the group of *Algæ*, to which *Protococcus* belongs, give rise, under certain circumstances, to locomotive bodies propelled by cilia, like the locomotive *Protococcus*, so that there is nothing anomalous in the case of Protococcus.

Like the yeast-plant, *Protococcus* retains its vitality after it has been dried. It has been preserved for as long as two years in the dry condition, and at the end of that time has resumed its full activity when placed in water. The wide distribution of *Protococcus* on the tops of houses and elsewhere, is thus readily accounted for by the transport of the dry *Protococci* by winds.

LABORATORY WORK.

A. MORPHOLOGY.

a. Resting or stationary Protococcus.

1. Spread out in water some mud from a gutter or
    similar locality, and put on a cover-glass. ' Look for
    the red or green protococcus cells with a low
    power. Having found some, put on a high power
    and make out the following points.

    Size ; (measure)—very variable.

    Form ; more or less spheroidal, with individual
    variations.

    Structure ; sac—protoplasm—sometimes a vacuole—
    sometimes apparently a nucleus. (Compare
    Torula, I. A. 2. *b.*)

    Colour ; generally green—sometimes red—sometimes
    half and half—sometimes centre red, periphery
    green—the colouring matter always in the pro-
    toplasm only—most frequently diffused, but
    sometimes in distinct granules, or oily looking
    drops.

2. Note especially the following forms of cell—

    a.   The primitive or normal form.

    Roundish cells, with a cellulose sac, and unseg-
    mented granular contents. Draw several carefully
    to scale. Apply the methods of mechanical and
    chemical analysis detailed for Torula. (I. A. 3. 4.
    5. 6.) Note that iodine in some cells produces a
    blue coloration by its action on the red matter
    present. Treat a specimen with strong iodine
    solution and then with sulphuric acid (75 per
    cent.) : the sac will become stained blue.

   b.   Cells multiplying by fission :

      α. *Simple fission.* The cell elongates, and the protoplasm divides into two across its longer axis, and then a partition is formed sub-dividing the sac; the halves either separate at once, and each rounds itself off and becomes an independent cell ; or one or both halves again divide, in .a similar way, before they separate, and so three or four new cells are produced.

      β. *Cells multiplying by budding, like Torula*; rare.

b. Motile stage.

      a. Mount a drop of water containing motile Protococcus, and examine with a high power. Note the actively locomotive green bodies, of which two varieties can be distinguished.

        α. Cells like the stationary ones in size, and apparently directly formed from them. Each possesses a structureless colourless sac, surrounding the coloured protoplasm, but the latter has shrunk away from the sac at most points.

          Note in various specimens—The two cilia prolonged from the protoplasm through apertures in the sac; their motionless part within the sac; their vibratile portion outside it. The colourless thin external layer of the protoplasm collected into a little heap at the point from whence the cilia arise. The delicate colourless processes radiating from the outer protoplasmic layer to the interior of the sac. The colour—usually green, but frequently one bright red spot is present.

β.   Cells much like the above if the cellulose sac were removed, and the radiating processes extending to it from the protoplasm withdrawn.

b.   Try to find specimens in which the movements are becoming sluggish, and see the cilia in motion.

c.   Stain with iodine: this kills the cells, and stops their movements; and frequently renders the cilia very distinct.

[B.   PHYSIOLOGY.

Get some water that is quite green from containing a large quantity of Protococcus; introduce some of it into two tubes filled with and inverted over mercury, and pass a small quantity of carbonic anhydride into each: keep one tube in the dark and place the other in bright sunlight for some hours.   Then measure the gas in each tube and afterwards introduce a fragment of caustic potash into each; the gas from the specimen kept in the dark will be more or less completely absorbed (= carbonic anhydride), that from the other will not be absorbed by the potash alone, but will be absorbed on the further introduction of a few drops of solution of pyrogallic acid (= oxygen).   Protococcus, therefore, in the sunlight, takes up carbonic anhydride and evolves oxygen.   A comparative experiment may be made with a third tube containing water but no Protococcus.]

III.

THE PROTEUS ANIMALCULE (*Amœba*). COLOURLESS BLOOD CORPUSCLES.

Amœbæ are minute organisms of very variable size which occur in stagnant water, in mud, and even in damp earth, and are frequently to be obtained by infusing any animal matter in water and allowing it to evaporate while exposed to direct sunlight.

An *Amœba* has the appearance of a particle of jelly, which is often more or less granular and fluid in its central parts, but usually becomes clear and transparent, and of a firmer consistency, towards its periphery. Sometimes *Amœbæ* are found having a spherical form and encased in a structureless sac, and in this encysted state they exhibit no movements. More commonly, they present incessant and frequently rapid changes of form, whence the name of "*Proteus Animalcule*" given to them by the older observers; and these changes of form are usually accompanied by a shifting of position, the *Amœba* creeping about with considerable activity, though with no constancy of direction.

The changes of form, and the movements, are effected by the thrusting out of lobe-like prolongations of the peripheral part of the body, which are termed *pseudopodia*, sometimes from one region and sometimes from another. Occasionally, a particular region of the body is constantly free from pseudopodia, and therefore forms its hindmost part when

M.

it moves. Each pseudopodium is evidently, at first, an extension of the denser clear substance (*ectosarc*) only; but as it enlarges, the central, granular, more fluid substance flows into its interior, often with a sudden rush.

In some *Amœba* a clear space makes its appearance, at intervals, in a particular region of the ectosarc, and then disappears by the rapid approach of its walls. After a while, a small clear speck appears at the same spot and slowly dilates until it attains its full size, when it again rapidly disappears as before. Sometimes two or three small clear spots arise close together, and run into one another to form the single large cavity. The structure thus described is termed the *contractile vesicle* or *vacuole*, and its rhythmical systole and diastole often succeed one another with great regularity. Nothing is certainly known respecting its function, nor even whether it does or does not communicate with the exterior, and thus pump water into and out of the body of the *Amœba*, though there is some reason to think that this may be the case.

Very frequently one part of the *Amœba* exhibits a rounded or oval body, which is termed the *nucleus*. This structure sometimes has a distinctly vesicular character, and contains a rounded granule called the *nucleolus*.

The gelatinous body of the *Amœba* is not bounded by anything that can be properly termed a membrane; all that can be said is, that its external or limitary layer is of a somewhat different constitution from the rest, so that it acquires a certain appearance of distinctness when it is acted upon by such reagents as acetic acid, or when the animal is killed by raising the temperature to 45°C. Physically, the ectosarc might be compared to the wall of a soap-bubble, which, though fluid, has a certain viscosity, which not only enables its particles to hold together and form a continuous sheet, but permits a rod to be passed into or through the bubble without bursting it; the walls closing together, and recovering their continuity, as soon as the rod is drawn away.

It is this property of the ectosarc of the *Amœba* which enables us to understand the way in which these animals take in and throw out again solid matter, though they have neither mouth, anus, nor alimentary canal. The solid body passes through the ectosarc, which immediately closes up and repairs the rent formed by its passage. In this manner, the *Amœbæ* take in the small, usually vegetable, organisms, which serve them for food, and subsequently get rid of the undigested solid parts.

The chemical composition of the bodies of the *Amœbæ* has not been accurately ascertained, but they undoubtedly consist, in great measure, of water containing a protein compound, and are similar to other forms of protoplasm. They absorb oxygen and give out carbonic acid, and the presence of free oxygen is necessary to their existence. When the medium in which they live is cooled down to the freezing point their movements are arrested, but they recover when the temperature is raised. At a temperature of about $35°$ C. their movements are arrested, and they pass into a condition of "heat-stiffening," from which they recover if that temperature is not continued too long; at $40°$ to $45°$ C. they are killed.

Electric shocks of moderate strength cause *Amœbæ* at once to assume a spherical still form, but they recover after a while. Strong shocks kill them.

Not unfrequently, an active Amœba becomes still spontaneously, acquires a rounded form, and secretes a structureless case or cyst, in which it remains enclosed for a shorter or longer period.

2—2

If *Amœba* are not to be found, their nature may be understood by the examination of bodies, in many respects very similar to them, which occur in the blood of all vertebrate and most invertebrate animals, and are known as the '*colourless corpuscles.*' They are to be met with in abundance in a fresh-drawn drop of human blood. In such a drop, after the red corpuscles have run into rolls, irregular bodies will be seen here and there in the meshes of the rolls. If one of these bodies is carefully watched it will be seen to undergo changes of form of the same character as those exhibited by *Amœba,* and these motions become much more active if the drop is kept at the temperature of the body by means of a hot stage. Each corpuscle is, in fact, a mass of protoplasm containing a nucleus, and the protoplasm sends out pseudopodia which are strictly comparable to those of *Amœba.* The colourless corpuscles, however, possess no contractile space.

The colourless corpuscles of the blood of some of the cold-blooded vertebrates, such as Frogs and Newts, may be kept alive for many weeks in serum properly protected from evaporation ; and if finely divided colouring matter, such as indigo, is supplied to them, either in the body or out of it, they take it into their interior in the same way as true Amœbæ would. In the earliest condition of the embryo, the whole body is composed of such nucleated cells as the colourless corpuscles of the blood ; and the colourless corpuscles must be regarded as simply the progeny of such cells, which have not become metamorphosed, and have retained the characteristics of the lowest and most rudimentary forms of animal life.

The *Amœba* is an animal, not because of its contractility or power of locomotion, but because it never becomes enclosed within a cellulose sac, and because it is devoid of the power of manufacturing protein from bodies of a comparatively simple chemical composition.   The *Amœba* has to obtain its protein ready made, in which respect it resembles all true animals, and therefore is, like them, in the long run, dependent for its existence upon some form or other of vegetable life.

LABORATORY WORK.

A. AMŒBA.

Place a drop of water containing *Amœbæ* on a slide, cover with a cover glass, avoiding pressure, and search over with ¼ inch obj.: having found an Amœba, examine with a higher power.

1. *Size:* differing considerably in different specimens. Measure.

2. *Outline:* irregular, produced into a number of thick rounded eminences (*pseudopodia*) which are constantly undergoing changes : sketch it at intervals of five seconds.

3. *Structure:*

    a. Outer hyaline border (*ectosarc*), tolerably sharply marked off: granular layer (*endosarc*) inside this, gradually passing into a more fluid central part.

    b. *Nucleus :* (absent in some specimens) ; a roundish more solid-looking particle, which does not change its form.

    c. *Contractile vesicle :* in the ectosarc note a roundish clear space which disappears periodically, and after a short time reappears; its slow diastole —rapid systole.   Not present in all specimens.

    d.   *Foreign bodies* (swallowed); Diatom cases, *Desmidiæ,* &c.

4. *Movements:*

    a.   Watch the process of formation of a *pseudopodium.* A hyaline elevation at first; then, as it increases in size, currents carrying granules flow into it.

    b.   Locomotion: watch the process,—a pseudopodium is thrown out, then the rest of the body appears to flow up to it, and the process is repeated.

    c.   If the opportunity presents itself, watch the process of the ingestion of solid matters.

    d.   [Observe the movements on the hot stage; warmth at first accelerates the movements, but as the temperature approaches 40° C. they cease, and the whole mass remains as a motionless sphere.]

    e.   [Effects of electrical shocks on the movements.]

5. *Mechanical Analysis:* crush. The whole collapses, except sometimes the nucleus, and even that after a time disappears: there is no trace of a distinct resisting outer sac.

6. *Chemical Analysis:* Treat with magenta and iodine. The whole stains, and there is no unstained enveloping sac. Iodine as a rule produces no blue coloration; when blue specks become visible it is probable that the starch which they indicate has been swallowed.

7. [Look for encysted specimens: and for specimens which are undergoing fission.]

8. Another form of Amœba is not unfrequently found which differs from that just described in being much less coarsely granular, and in having no well-defined ectosarc and endosarc, and also in having much longer, more slender and pointed pseudopodia. Another common form progresses rapidly with a slug-like movement, only throwing out pseudopodia at its anterior end.

B. WHITE BLOOD-CORPUSCLES, (human).

Prick your finger and press out a drop of blood : spread out on a slide under a coverslip, avoiding pressure, and surround the margin of the coverglass with oil. Neglect the pale yellow homogeneous (*red*) corpuscles, and examine the much less numerous, granular, colourless, ones.

Note their—

1. *Size:* (measure).

2. *Form:* changing much like that of the Amœba, but less actively.   Draw at intervals of ten seconds.

3. *Structure:* Some more and some less granular; but no distinct ectosarc, endosarc, and vacuole as in the Amœba. Nucleus rarely visible in the fresh state. No contractile vesicle.

4. Treat with dilute acetic acid : the granules are cleared up, and a nucleus is brought into view in a more or less central position. If the acetic acid has been too strong the nucleus will be constricted and otherwise distorted.

5. Stain with magenta, and iodine; the whole becomes coloured, the nucleus most intensely.

6. Place on the hot stage, and gradually warm up to 50° C. The movements are at first rendered more

active, but ultimately cease, the pseudopodia-like processes being all retracted and the whole forming a motionless sphere.

Let the specimen cool again; the movements are not resumed; the protoplasm having passed into a state of permanent coagulation or rigidity.

7. Repeat the above observations on the white blood-corpuscles of the frog or newt.

IV.

BACTERIA.

UNDER the general title of *Bacterium* a considerable variety of organisms, for the most part of extreme minuteness, are included.

They may be defined as globular, oblong, rod-like or spirally coiled masses of protoplasmic matter enclosed in a more or less distinct structureless substance, devoid of chlorophyll and multiplying by transverse division. The smallest are not more than $\frac{1}{30000}$th of an inch in diameter, so that under the best microscopes they appear as little more than mere specks, and even the largest have a thickness of little more than $\frac{1}{10000}$th of an inch, though they may be very long in proportion. Many of them have, like *Protococcus*, two conditions—a still and an active state. In their still condition, however, they very generally exhibit that *Brownian* movement which is common to almost all very finely divided solids suspended in a fluid. But this motion is merely oscillatory, and is readily distinguishable from the rapid translation from place to place which is effected by the really active *Bacteria*.

In one of the largest forms, *Spirillum volutans*, it has been possible to observe the cilia by which the movement is effected. In this there is a cilium at each end of the spirally coiled body. No such structure, however, can be made out in the straight *Bacteria*, and it remains doubtful whether they possess cilia which are too fine to be rendered

visible by our microscopes, or whether their movements are
due to some other cause. Many forms, such as the *Vibriones*,
so common in putrefying matters, appear obviously to have
a wriggling or serpentiform motion, but this is an optical
illusion. In this Bacterium, as in all others, the body does
not rapidly change its form ; but its joints are bent zig-zag-
wise, and the rotation of the zig-zag upon its axis, as it
moves, gives rise to the appearance of undulatory contraction. A cork-screw turned round, while its point rests
against the finger, gives rise to just the same appearance.

Bacteria, in the still state, very often become surrounded
by a gelatinous matter, which seems to be thrown out by
their protoplasmic bodies, and to answer to the cell-wall of
the resting *Protococcus*. This is termed the *Zoogloea* form of
Bacterium.

Bacteria grow and multiply in Pasteur's solution (without sugar) with extreme rapidity, and, as they increase in
number, they render the fluid milky and opaque. Their
vital actions are arrested at the freezing point. They thrive
best in a temperature of about 30° C. but, in most fluids,
they are killed by a temperature of 60° C. (140° F.).

In all these respects *Bacteria* closely resemble *Torula;*
and a further point of resemblance lies in the circumstance
that many of them excite specific fermentative changes in
substances contained in the fluid in which they live, just as
yeast excites such changes in sugar.

All the forms of putrefaction which are undergone by
animal and vegetable matters are fermentations set up by
Bacteria of different kinds. Organic matters freely exposed
to the air are, in themselves, nowise unstable bodies, and,
if due precautions have been taken to exclude *Bacteria*,
they do not putrefy, so that, as has been well remarked,
"putrefaction is a concomitant not of death, but of life."

Bacteria, like *Torulæ* and *Protococci*, are not killed by drying up, and from their excessive minuteness they must be carried about still more easily than *Torulæ* are. In fact there is reason to believe that they are very widely diffused through the air, and that they exist in abundance in all ordinary water and on the surface of all vessels that are not chemically clean. They may be readily filtered off from the air, however, by causing it to pass through cotton wool.

LABORATORY WORK.

1. Infuse some hay in warm water for half an hour—filter, and set aside the filtrate: note the changes which go on in it—at first clear, in 24 or 36 hours it becomes turbid; later on, a scum forms on the surface and the infusion acquires a putrefactive odour.

2. Rub some gamboge down in water and examine a drop of the mixture with a high power: avoid all currents in the fluid and watch the *Brownian movements;* note that they are simply oscillatory—not translative.

3. Take a drop of fluid from a turbid hay infusion—and examine it, using the highest power you have; in it will be found multitudes of

Moving Bacteria. Note their—

    a. *Form;* elliptic or rodlike—sometimes forming short (2—8) jointed rows.

    b. *Size;* breadth, very small but pretty constant; length, varying, but several times greater than their breadth: measure.

c. *Structure;* an outer more transparent layer enveloping less transparent matter: in the compound forms the envelope appears only where two joints come in contact, so that the rod looks as if made up of alternating transparent and more opaque substances.

d. *Movements;* some vital, and some purely physical (*Brownian*). The former various but progressive: the latter, a rotatory movement round a stationary centre; study it in a drop of boiled infusion in which the Bacteria are all dead.

4. Treat with iodine—only the more opaque parts stain; probably then we have to do with protoplasm, enveloped in nonprotoplasmic matter.

5. **Resting Bacteria.** (*Zoogloea-stage.*)

   a. Examine the scum from the surface of a hay infusion; it exhibits myriads of motionless Bacteria, embedded in gelatinous material.

   b. Treat with iodine; the Bacteria stain as before : the gelatinous uniting material remains unstained.

6. Mixed with the Bacteria proper, both in the pellicle and the fluid beneath, may be found the following forms of living beings.

   a. **Micrococcus.** Bodies much like Bacteria, but short and rounded, and occurring singly, or in bead-like rows. They may be found free or in a Zoogloea stage.

   b. **Bacillus.** Threads composed of straight cylindrical joints much longer than those of Bacteria but of a similar structure: they are always free-swimming.

c. **Vibrio.** Like Bacillus, but with bent joints.

d. **Spirillum.** Elongated unjointed threads rolled up into a more or less perfect spiral: frequently two spirals intertwine. In some of the largest forms a vibratile cilium can be made out on each end of the thread.

e. **Spirochæte.** Much like spirillum, but longer and with a much more closely rolled spiral. A very actively motile but not common form.

7. Examine various putrefying fluids for Bacteria and related organisms.

8. Place some fresh-made hay infusion in three flasks; boil two of them for three or four minutes, and while one is boiling briskly stop its neck with a plug of cotton-wool and continue to boil for a minute or two: leave the necks of the other two flasks unclosed, and put all three away in a warm place.

a. In a day or two abundant Bacteria will be found in the unboiled flask.

b. In the boiled but unclosed flask Bacteria will also appear, but perhaps not quite so soon as in *a.*

c. In the flask which has been boiled and kept closed Bacteria will not appear, if the experiment has been properly performed, even if it be kept for many months.

V.

MOULDS (*Penicillium and Mucor*).

Torula, Protococcus and *Amœba* are extremely simple conditions of the two great kinds of living matter which are
known as Plants and Animals. No plants are simpler in
structure than *Torula* and *Protococcus*, and the only animals which are simpler than *Amœba*, are essentially *Amœbæ*
devoid of a nucleus and contractile vesicle. Moreover, however complicated in structure one of the higher plants may
be in its adult state, when it commences its existence it is as
simple as *Torula* or *Protococcus*, or at most as *Torula* or *Protococcus* would be if it possessed a distinct nucleus; and the
whole plant is built up by the fissive multiplication of the
simple cell in which it takes its origin, and by the subsequent growth and metamorphosis of the cells thus produced.
The like is true of all the higher animals. They commence
as nucleated cells, essentially similar to Amœbæ and colourless blood-corpuscles, and their bodies are constructed by
aggregations of metamorphosed cells, produced by division
from the primary cell. It has been seen that *Torula*
and *Protococcus*, similar as they are in structure, are distinguished by certain important physiological peculiarities;
and the more complicated plants are divisible into two
series, one produced by the growth and modification of cells
which have the physiological peculiarities of *Torula* and
contain no chlorophyll, while the other, and far larger, series

presents chlorophyll, and has the physiological peculiarities
of *Protococcus.* The former series comprises the *Fungi,* the
latter all other plants; only a few parasitic forms among
these being devoid of chlorophyll.

The *Fungi* take their origin in *spores,* a kind of cells,
which, however much they may vary in the details of their
structure, are essentially similar to *Torulæ.* Indirectly or
directly, the spore gives rise to a long tubular filament,
which is termed a *hypha,* and out of these hyphæ the
Fungus is built up.

One of the commonest Moulds, the *Penicillium glaucum,*
which is familiar to every one from its forming sage-green
crusts upon bread, jam, old boots, &c. affords an excellent
and easily studied example of a Fungus. When examined
with a magnifying glass, the green appearance is seen to be
due, in great measure, to a very fine powder which is detached from the surface of the mould by the slightest touch.
Beneath this lies a felt-work of delicate tubular filaments,
the hyphæ, forming a crust like so much blotting-paper,
which is the *mycelium.* From the free surface of the crust
innumerable hyphæ project into the air and bear the green
powder. These are the *aerial hyphæ.* On the other hand, the
attached surface gives rise to a like multitude of longer
branched hyphæ, which project into the fluid in which the
crust is growing, like so many roots, and may be called the
submerged hyphæ. If the patch of *Penicillium* has but a
small extent relatively to the surface on which it lies, multitudes of silvery hyphæ will be seen radiating from its
periphery and giving off many submerged, but few or no
vertical, or subaërial, branches. Submitted to microscopic
examination, a hypha is seen to be composed of a transparent wall (which has the same characters as the cell-wall of
Torula) and protoplasmic contents, which fill the tube

formed by the wall, and present large central clear spaces, or vacuoles. At intervals, transverse partitions, continuous with the walls of the tube, divide it into elongated cells, each of which contains a correspondingly elongated protoplasmic sac, or *primordial utricle.* The hyphæ frequently branch dichotomously; and, in the crust, they are inextricably entangled with one another; but every hypha, with its branches, is quite distinct from every other. Those aerial hyphæ which are nearest the periphery of the crust end in simple rounded extremities; but the others terminate in brushes of short branches, and each of these branches, as it grows and elongates, becomes divided by transverse constrictions into a series of rounded spores arranged like a row of beads. The spores formed in this manner are termed *conidia.* At the free end of each filament of the brush the conidia become very loosely adherent, and constitute the green powdery matter to which reference has been made. Examined separately, a *conidium* is seen to be a spherical body, composed of a transparent sac, enclosing a minute mass of protoplasm, in all essential respects similar to a *Torula.* If sown in an appropriate medium, as for example Pasteur's solution, with or without sugar, the *conidium* germinates. Upon from one to four points of its surface an elevation or bulging of the cell-wall and of its contained protoplasm appears. This rapidly increases in length, and, continually growing at its free end, gives rise to a hypha, so that the young *Penicillium* assumes the form of a star, each ray being a hypha. The hyphæ elongate, while side branches are developed from them by outgrowths of their walls; and this process is repeated by the branches, until the hyphæ proceeding from a single conidium may cover a wide circular area, as a patch of mycelium. When, as is usually the case, many conidia germinate close together, their hyphæ cross one another, interlace, and give rise to a papyraceous crust. After the hyphæ have attained a certain length, the protoplasm divides at intervals, and transverse septa are formed between the masses thus divided off from one another. But neither in this, nor in any other Fungus, are septa formed in the direction of the length of the hypha.

Very early in the course of the development of the mycelium, branches of the hyphæ extend downwards into the medium on which the mycelium grows; while, as soon as the patch has attained a certain size, the hyphæ in its centre give off vertical aerial branches, and the development of these goes on, extending from the centre to the periphery. The outgrowth of pencil-like bunches of branches at the end of these takes place in the same order; and these branches, becoming transversely constricted as fast as they are formed, break up into conidia, which are ready to go through the same course of development.

The conidia may be kept for a very long time in the dry state, without their readiness to germinate being in any way impaired, and their extreme minuteness and levity enable them to be dispersed and carried about by the slightest currents of air. The persistence of their vitality is subject to nearly the same conditions of temperature as that of yeast. Not unfrequently *Torulæ* make their appearance, in abundance, among the hyphæ and conidia of *Penicillium*, and appear to be derived from them; but it is still a disputed point, whether they are so or not.

If some fresh horse-dung be placed in a jar and kept moderately warm, its surface will, in two or three days, be covered with white cottony filaments, many of which rise vertically into the air, and end in rounded heads, so that they somewhat resemble long pins. The organism thus produced is another of the Fungi—the mould termed *Mucor muuedo*.

Each rounded head is a *sporangium ;* the stalk on which it is supported rises from one of the filaments which ramify in the substance of the horse-dung, and are the *hyphæ*. Each hypha is, as in *Penicillium*, a tube provided with a tough thickish structureless wall, which is partly composed of cellulose, and is filled by a vacuolated protoplasm. In old specimens, transverse partitions, continuous with the walls of the hyphæ, may divide them into chambers or cells. The stalk of the sporangium is a hypha of the same structure as the others. The wall of the sporangium is beset with minute asperities composed of oxalate of lime, and it contains a great number of minute oval bodies, the *spores*, held together by a transparent intermediate substance. When the sporangium is ripe, the slightest pressure causes its thin and brittle coat to give way, and the spores are separated by the expansion of the intermediate substance, which readily swells up and finally dissolves, in water. The greater part of the wall of the sporangium then disappears, but a little collar, representing the remains of its basal part, frequently adheres to the stalk. The cavity of the stalk does not communicate with that of the sporangium, but is separated from it by a partition, which bulges into the cavity of the sporangium, forming a central pillar or projection. This is termed the *columella* and stands conspicuously above the collar, when the sporangium has burst and the spores are evacuated.

The spores are oval and consist of a sac, having the same composition as the wall of the hypha, which encloses a mass of protoplasm. When they are sown in an appropriate medium, as for example in Pasteur's solution, they enlarge, become spheroidal, and then send out several thick prolongations. Each of these elongates, by constant growth at its free end, and becomes a hypha, from which branches are given off, which grow and ramify in the same way. As all the ramifying hyphæ proceed from the spore as a centre, their development gives rise, as in *Penicillium*, to a delicate stellate *mycelium*. At first, no septa are developed in the hyphæ, so that the whole mycelium may be regarded as a single cell with long and ramified processes, and the *Mucor*, at this stage, is an unicellular organism. From near the centre of the mycelium a branch is given off from a hypha, rises vertically, and after attaining a certain length ceases to elongate. Its free end dilates into a rounded head, which gradually increases in size, until it attains the dimensions of a full-grown sporangium; and, at the same time, the protoplasm contained in this head becomes separated from that in the stalk by a septum, which is curved towards the cavity of the sporangium, and constitutes the columella. The wall of the sporangium, thus formed, becomes covered externally with a coat of oxalate of lime spines. As the sporangium increases in size, its protoplasmic contents become marked out into a large number of small oval masses, which are close together, but not in actual contact. Each of these masses next becomes completely separate from the rest, surrounds itself with a cellulose coat, and becomes a spore; while the protoplasm not thus used up in the formation of spores, appears to give rise to the gelatinous intermediate substance, which swells up in water, referred to above. The walls of the spores become coloured, and that of the sporangium gradually thins, until it is reduced to little more than the outer crust of oxalate of lime. The sporangium now readily bursts, and the spores are separated by the

3—2

swelling and eventual dissolution of the gelatinous intermediate matter. Sporangia, in which spores are produced by
division of the protaplasm, are commonly termed *asci*, and
the spores receive the name of *ascospores*.

There appears to be no limit to the extent to which the
Mucor may be reproduced by this process of *asexual* development of spores, by the fission of the contents of the
sporangium; nor does any other mode of multiplication
become apparent, so long as the mould grows in a fluid
medium and is abundantly supplied with nourishment.

But when growing in nature, in such matters as horse-
dung, a method of reproduction is set up which represents
the sexual process in its simplest form. Adjacent hyphæ,
or parts of the same hypha, give off short branches, which
become dilated at their free ends, and approach one another, until these ends are applied together. The protoplasm in each of the dilated ends becomes separated by a
septum from that of the rest of the branch; the two cells
thus formed open into one another by their applied faces,
and their protoplasmic contents becoming mixed together,
form one spheroidal mass, to the shape of which the coalesced cell-membranes adapt themselves. This process of
conjugation evidently represents that of sexual impregnation
among higher organisms, but as there is no morphological
difference between the modified hyphæ which enter into
relation with one another, it is impossible to say which
represents the male, and which the female, element. The
product of conjugation is termed a *zygospore*. Its cellulose
coat becomes separated into an outer layer of a dark black-
ish hue, the *exosporium*, and an inner colourless layer, the
endosporium. The outer coat is raised into irregular elevations, to which corresponding elevations of the inner coat
correspond.

Placed in favourable circumstances, the *zygospore* does not immediately germinate; but, after a longer or shorter period of rest, the exosporium and the endosporium burst, and a bud-like process is thrown out, which, usually, grows only into a very short unbranched hypha. From this hypha a vertical prolongation is developed, which becomes converted into a sporangium, such as that already described, whence spores are produced, which give rise to the ordinary stellate mycelium. Thus, *Mucor* presents what is termed an "*alternation of generations*." The zygospore resulting from a sexual process developes into a rudimentary mycelium, with a single sporangium which constitutes the first generation (*A*). This gives rise, by the asexual development of spores in its sporangium, to the second generation (*B*), represented by as many separate *Mucores* as there are spores. The second generation (*B*) may give rise sexually to zygospores and so reproduce the generation (*A*); but, more usually, an indefinite series of generations similar to (*B*) are produced from one another asexually, before (*A*) returns.

When *Mucor* is allowed to grow freely at the surface of a saccharine liquid, it takes on no other form than that described; but, if it be submerged in the same liquid, the mode of development of the younger hyphæ becomes changed. They break up, by a process of constriction, into short lengths, which separate, acquire rounded forms, and at the same time multiply by budding after the manner of *Torulæ*. Coincidentally with these changes, an active fermentation is excited in the fluid, so that this "*Mucor-Torula*," functionally as well as morphologically, deserves the name of 'yeast.'

If the *Mucor-Torula* is filtered off from the saccharine solution, washed, and left to itself in moist air, the *Torulæ* give off very short aërial hyphæ, which terminate in minute sporangia. In these a very small number of ordinary mucor spores is developed, but, in essential structure, both the sporangia and the spores resemble those of normal *Mucor*.

LABORATORY WORK.

A. PENICILLIUM.

Prepare some Pasteur's fluid, and leave it exposed to the air in saucers in a warm place; if *Penicillium* spores are at hand add a few to the fluid in each saucer: if spores cannot be obtained, the fluid, if simply left to itself, will probably be covered with *Penicillium* in ten days or a fortnight. Sometimes, however, the fluid will overrun with *Bacteria*, to the exclusion of everything else. And very frequently other moulds, such as *Aspergillus*, or *Mucor*, may appear instead of or along with *Penicillium*.

1. NAKED-EYE CHARACTERS. Note the powdery-looking upper surface, white in young specimens, pale greenish in older, and later still becoming dark sage-green: the smooth pale under surface: the dense tough character of the mycelium.

2. HISTOLOGICAL STRUCTURE.

   a. **The mycelium.**

      a. Tease a bit out in water, and examine first with low, and then with a high power: it is chiefly made up of interlaced threads or tubes—the

        α. *Hyphæ.* Note their diameter (measure)—form—subdivisions (*cells*)—dichotomous mode of branching—and structure: the external

homogeneous sac; the granular less transparent protoplasm; the small round vacuoles.
Draw.

β. *The intermixed Torulæ.* Note their size and
number.

b. Hold a bit of the mycelium between two pieces
of carrot, and cut a thin vertical section with a
sharp razor: mount in water and examine with
low and high power.

b. **The submerged hyphæ.**

Small branched threads hanging down from the under
surface of the mycelium: repeat the observations
2. a. a. a.

c. **The aërial hyphæ and conidiophores.**

Tease out in water a bit from the surface of one of
the greenish patches; observe the difficulty with
which water wets it. Examine with low and high
power.

Note ;—

α. The primary erect hypha.

β. Its division into a number of branches.

γ. The division of the terminal branches by constrictions into a chain of conidia. Draw.

d. **The conidia.**

a. Their *Size* (measure). .

Form; spherical.

Structure; sac, protoplasm, vacuole.

b. Stain with magenta and iodine.

c. Treat another specimen with potash.

e. The germination of the Conidia, and building up of the Mycelium.

a. Sow some conidia in Pasteur's fluid in a watch-glass; protect from evaporation, and watch the development of the mycelium (examine the surface with a low power); then the formation of aërial hyphæ; finally the production of new conidia.

b. [Sow Conidia in Pasteur's fluid in a moist chamber, and watch from day to day; note the formation of eminences at one or more points on a conidium; the elongation of these eminences to form hyphæ; the branching and interlacement of the hyphæ.]

B. MUCOR MUCEDO.

1. Place some fresh horse-dung under a bell-jar and keep moist and warm; in from 24 to 48 hours its surface will nearly always be covered by a crop of erect aërial mucor-hyphæ, each ending in a minute enlargement (*sporangium*) just visible with the unassisted eye: it is this first crop of hyphæ and sporanges which is to be examined.

2. Snip off a few of the hyphæ with a pair of scissors, mount in water, and examine with 1 inch obj.

   a. Large unbranched hyphæ, each ending in a spherical enlargement (*sporangium*).

3. Examine with $\frac{1}{8}$ obj.

   a. The hyphæ.

      α. *Their size;* they greatly exceed the hyphæ of *Penicillium* both in length and diameter.

      β. *Their structure;* homogeneous sac, granular protoplasm, vacuoles: septa absent except close to the sporange.

γ. Treat with iodine and magenta; the proto-
    plasm is stained.

δ. Treat another specimen with Schulz's solu-
    tion; the wall is stained violet.

b. **The sporangia or asci.**
    Examine with ⅛ obj.

  a. Their size and *form.*
  b. Their structure.

    α. The homogeneous enveloping sac covered by
        irregular masses of calcic oxalate.

    β. The granular protoplasmic contents: un-
        segmented in some; divided into a great
        number of distinct oval masses (*ascospores*) in
        others.

    γ. The projection into the sporangial cavity of
        the convex septum (*columella*) which separates
        the hypha from the sporange.

    δ. The *collar* projecting around the base of the
        columella of burst sporangia.

  *c. Stain some with iodine; others with Schulz's
       solution.

c. **The ascospores.**

  *a. Crush some ripe asci by gentle pressure upon
       the cover-glass.   Examine with ⅛ obj.

    α. The size of the ascospores (*measure*).
    β. Their form; cylindrical and elongated.
    γ. Their structure.
    δ. Stain with iodine and magenta.

VI.

STONEWORTS (*Chara and Nitella*).

THESE water-weeds are not uncommonly found in ponds and rivers, growing in tangled masses of a dull green colour. Each · plant is hardly thicker than a stout needle, but may attain a length of three or four feet. One end of the *stem* is fixed in the mud at the bottom, by slender thread-like roots, the other floats at the surface. At intervals, *appendages*, consisting of *leaves, branches, root-filaments*, and *reproductive organs*, are disposed in circles, or *whorls*. In the middle and lower parts of the plant these whorls are disposed at considerable and nearly equal distances; but, towards the free upper end, the intervals between the whorls diminish, and the whorled appendages themselves become shorter, until, at the very summit, they are all crowded together into a terminal bud, which requires the aid of the microscope for its analysis.

The parts of the stem, or *axis*, from which the appendages spring are termed *nodes;* the intervening parts being *internodes*. When viewed with a hand-magnifier the internodes exhibit a spiral striation.

In *Chara*, each internode consists of a single, much-elongated cell, which extends throughout its whole length, invested by a *cortical layer*, composed of many cells, the spiral arrangement of which gives rise to the superficial marking which has been noted. And this multicellular structure is continued from the cortical layer, across the

stem, at each node. The stem therefore consists of a series
of long, axial cells, contained in as many closed chambers
formed by the small cortical cells. The nodes are the multicellular partitions between these chambers. The branches
are altogether similar in structure to the main stem. The
leaves are also similar to the stem, so far as they consist of
axial and cortical cells, but they differ in the form and
proportions of these cells, as well as in the fact that the
summit, or free end, of the leaf is always a much-elongated
pointed cell. The branches spring from the re-entering
angle between the stem and the leaf, which is termed the
axilla of the leaf; and, in the same position, at the fruiting
season of the plant are found the reproductive organs.
These are of two kinds, the one large and oval, the *sporangia*
or *spore-fruits*, the other smaller and globular, the *antheridia*.
Both, when ripe, have an orange-red colour, and are seated
upon a short stalk.

If a growing plant be watched, it will be found that it
constantly increases in length two ways. New nodes, internodes, and whorls of appendages are constantly becoming
obvious at the base of the terminal bud ; and these appendages increase in size and become more and more widely
separated, until they are as large and as far apart as in the
oldest parts of the plant. The appendages at first consist
exclusively of leaves and root-filaments (*rhizoids*), and it is
only when these have attained their full size, that branches,
spore-fruits and antheridia are developed in their axillæ.
Sometimes rounded cellular masses appear in the axillæ of
the leaves, and, becoming detached, grow into new plants.
These are comparable to the *bulbs* of higher plants.

If the innermost part of the terminal bud, which constitutes the free end of the axis, or stem, be examined, it
will be found to be formed by a single nucleated cell,

separated by a transverse septum from another. Beneath this last follows another cell, which has already undergone division into several smaller cells by the development of longitudinal septa. This is the most newly-formed node. Below this again is a single cell, which is both longer and broader than those at the apex, and is an internodal cell. Below it follows another node, composed of more numerous small cells than in the first. Some of the peripheral cells of this node are undergoing growth and division, and thus give rise to cellular prominences, which are rudiments of the first whorl of leaves. In the still lower parts of the stem the internodal cells get longer and longer, but they never divide. The nodal cells, on the other hand, multiply by division, but do not greatly elongate. From the first, the nodal cells overlap the internodal cell, so as to meet round its equator, and thus completely invest it externally. And, as the internodal cell grows and elongates, the overlapping parts of the nodes increase in length and become divided into internodal and nodal cells, which take on a spiral arrangement, and thus give rise to the cortical layer.

Thus the whole plant is composed of an aggregation of simple cells; and, while it lives, new nodes and internodes are continually being added at its summit, or *growing point.* The internodal cells which give rise to the centre of the stem undergo no important change, except great increase of size, after they are once formed. The nodal cells, on the contrary, undergo division with comparatively little increase in size. And out of them, the nodes, the cortical layer, and all the appendages, are developed.

In all the young cells of *Chara* a *nucleus* of relatively large size is to be seen imbedded in the centre of the protoplasm, which is motionless, and is enclosed in a structureless cell-wall, containing cellulose. As the cell grows

larger, the centre of the protoplasm becomes occupied by a
watery fluid, and its thick periphery, which remains applied
against the cell-wall, constitutes the wall of a sac, or *primordial utricle*, in which the nucleus is imbedded.   In the
larger cells the primordial utricle is readily detached and
made to shrivel up into the middle of the cell by treatment
with strong alcohol.

  Numerous small green bodies—*chlorophyll grains*—are
imbedded in the outer, or superficial, part of the primordial
utricle.   And they increase in number by division, as the
cell enlarges.   These chlorophyll grains are composed of
protoplasmic matter, which frequently contains starch granules, and is impregnated with the green colouring substance.

  During life, the layer of the primordial utricle which
lies next to the watery contents of all the larger cells is in
a state of incessant rotatory motion, while the outermost
layer which contains the chlorophyll grains is quite still.
In the large cells, so long as the nucleus is discernible, it is
carried round with the rotating stream.

  The *antheridium* is a globular spheroidal body with a
thick wall, made up of eight pieces, which are united by
interlocking edges.   The four pieces which make up the
hemisphere to which the stalk of the antheridium is attached, are foursided, the other four are triangular.   From
the centre of the inner, concave face of each piece a sort of
short process, the handle or *manubrium*, projects into the
cavity of the hollow sphere.   At the free end of the manubrium is a rounded body, the *capitulum*, which bears six
smaller, *secondary capitula;* and each secondary capitulum
gives attachment to four long filaments divided by transverse partitions into a multitude (100 to 200) of small
chambers.   Thus, there may be as many as 20,000 to

40,000 chambers in each *antheridium* (8 × 6 × 4 × 100 or
× 200). The several pieces of which the wall of the antheridium is composed, the manubrium, the capitula, the
secondary capitula and the chambers of the filaments, are
all more or less modified cells, as may be proved by tracing
the antheridia from their earliest condition, as small processes of the nodal region, to their complete form. The
cells of the filaments are, at first, like any other cells; but,
by degrees, the protoplasm of each becomes changed into a
thread-like body, thicker at one end than at the other, and
coiled spirally like a corkscrew. From the thin end two
long cilia proceed; and, when the cells are burst, and the
antherozooids are set free, they are propelled rapidly, with
the small end forwards, by the vibration of the cilia. These
antherozooids answer to the spermatozoa of animals, and
represent the male element of the *Chara.*

The *sporangia* or *spore-fruits* are borne upon short stalks,
the end of which supports a large oval central cell; five
spirally-disposed sets of cells invest this, an aperture being
left between the investing cells at the apex of the sporangium. When the antheridia attain maturity they burst, the
antherozooids are set free, and swarm about in the water.
Some of them enter the aperture of the sporangium, and, in
all probability, pierce the free summit of the oval central cell,
and enter its protoplasm; but all the steps of this process
of impregnation have not been worked out. The result,
however, is, that the contents of the central cell become
full of starchy and oily matter; the spiral cells forming its
coat acquire a dark colour and hard texture, and the sporangium, detaching itself, falls into the mud.

After a time it germinates; a tubular process, like a
hypha, protrudes from its open end, and almost immediately
gives off a branch, which is the first root (compare the ger-

mination of the spore of a fern below). The hypha-like tube elongates, and becomes divided transversely into cells, the protoplasm of which developes chlorophyll. Very soon, the further growth of this *pro-embryo* is arrested. But one of the cells, which lies at some distance below the free end of the pro-embryo, undergoes budding, and gives rise to a set of leaves (which are not arranged in a whorl), amidst which a bud appears, which has the structure of the terminal bud of the adult *Chara* stem, and grows up into a new *Chara.*

We have then, in *Chara,* a plant which is *acrogenous* (or grows at its summit), and which becomes segmented by the development of appendages, at intervals, along an axis; which multiplies, asexually by bulb-like buds, and also multiplies sexually by means of the antherozooids (male elements) and central cells of the sporangia (female elements); in which the first product of the germination of the impregnated ovicell is a hypha-like body, from which the young *Chara* is developed by the germination and growth of one cell; so that there is a sort of *alternation of generations,* though the alternating forms are not absolutely distinct from one another.

Chara flourishes in pond-water under the influence of sunlight, and by the aid of its chlorophyll, so that its nutritive processes must be the same as those of *Protococcus.* From its complete immersion, and the absence of any duct-like, or vascular tissues, it is probable that all parts absorb and assimilate the nutriment contained in the water; and that, except so far as the reproductive organs are concerned, there is a morphological differentiation of organs, unaccompanied by a corresponding physiological differentiation.

Nitella is a rarer plant than *Chara,* and is simpler in structure, its axis being devoid of the cortical layer. In other respects, however, it is very similar to *Chara*, and its structure is more easily made out.

[The *Characeæ*, or plants belonging to the genera *Chara* and *Nitella*, are found in all parts of the world, and are in many respects closely allied to the *Algæ*, or water-weeds. But no *Algæ* are provided with an axis and appendages possessing a similar structure, or following the same law of growth, nor have any similar reproductive organs. The antherozooids of the *Characeæ* are, in fact, similar to those of the mosses, from which however the *Characeæ* differ widely in all other respects.]

LABORATORY WORK.

A. NAKED-EYE CHARACTERS.

Note the slender elongated axis (*stem*); the whorled appendages (*leaves*); the *nodes* and *internodes;* the shortening of the latter towards the apex of the stem; the *rhizoids*.

> a.   *The roots;* small; serving chiefly for attachment, the plant getting most of its nutrition, through other parts, from matters dissolved in the water.
>
> b.   *The leaves;* their sub-divisions (*leaflets*); their form, size, &c.
>
> c.   *The spore-fruits* and *antheridia;* their position, size, form, colour.

Draw a portion including two or three internodes.

B. HISTOLOGICAL STRUCTURE.

a. The stem.

1.   Examine the outside of a fresh internode with a low power, or pocket lens, to see the spirally-arranged cortical cells.

2. Hold a bit of fresh stem between two pieces of carrot, or imbed it in paraffin, and, with a sharp razor, cut thin transverse and longitudinal slices through nodes and internodes. Note the cavity of the large central cell (*medullary* or *internodal cell*) in the internodes; the *cortical cells*, round the medullary cell; the *nodal cells*, and the interruption of the central cavity at the nodes.

3. Examine similar sections in specimens treated with spirit, and also preparations made by teasing or pressing out in glycerine bits of stem from chromic acid (0·2 per cent.) preparations: make out in these,—

   α. The nodal, internodal, and cortical cells.

   β. The wall (*sac*), protoplasmic layer (*primordial utricle*), nucleus, and vacuole of each cell. (The nucleus is not always to be found in old cells.)

4. Examine sections from the fresh stem to make out the points detailed in B. a. 3. β. The protoplasm and nucleus are difficult to see. Note the chlorophyll-granules. (See B. b. γ.)

5. Stain sections of the fresh stem with iodine, and magenta: note the results.

   b. **The leaves.**

Examine fresh and chromic acid specimens.

   α. The large uncovered terminal cell.

   β. Then a series of internodal cells, separated from one another, and covered-in, by nodal cells: the sac, protoplasm, nucleus, and vacuole of each.

   γ. The *chlorophyll:* collected into oval granules, and arranged so as to leave an oblique uncoloured band round each cell; the position of these granules, in the more superficial layer of the protoplasm.

    δ.   The protoplasmic movements (see C. *a.*).

c.　The terminal bud.

Dissect out chromic acid specimens as far as possible with needles, and then press gently out in glycerine.　Note in different specimens—

a.　The terminal or apical cell:

    α.   Its *form:* hemispherical, the rounded surface free; the flat surface attached to the cell below it.

    β.   *Structure:* sac, protoplasm, nucleus; no vacuole present.

    γ.   Sometimes two nuclei; preliminary to division.

    δ.   Its mode of division; across the long axis of the stem, giving rise to two superimposed nucleated cells.

b.　The further fate of the new cells which are successively segmented off from the terminal cell; work back in your specimens from the terminal cell.

    α.   The new cells are successively *nodal* and *internodal;* the latter enlarge, develope a large vacuole, and ultimately form the medullary cells of the internodes; they never divide.

    β.   The nodal cells divide freely, and do not increase much in size; they give origin to the nodes and the cortical cells.

c.　The development of leaves: by the multiplication and outgrowth of nodal cells.

d. Their growth at the base, the terminal leaf-cell soon attaining its full size and not dividing.

e. *The development of branches;* from nodal cells in leaf-axils, which take on the character of terminal cells.

d. The spore-fruits.

Examine fresh, under a low power.

α. Made up externally of five twisted cells, bearing at their apices five smaller, not twisted cells.

β. Cut sections from imbedded specimens, and examine with a high power: make out the large central nucleated cell; the fatty and starchy matters contained in it; stain with iodine.

γ. Press out chromic acid specimens in glycerine; make out the above points (**d** *α, β*).

δ. Examine chromic acid specimens for young spore-fruits, and press them out in glycerine: make out in the youngest the five roundish cells surrounding a central one; then in older specimens the elongation, and twisting of the external cells, and the separation of their apices as five distinct cells.

e. The antheridia.

a. Examine, with a low power, a ripe (orange-coloured) one.

α. Make out its external dentated cells.

β. Tease out a ripe antheridium in water; and examine with a high power; note the flat, dentated, nucleated external cells; the cylindrical cell (*manubrium*) springing perpendicularly from the inner surface of each; the roundish cell (*capitulum*) on the inner end of the manubrium; the six *secondary capitula* attached to the capitulum; the thread-like filaments (usually four) proceeding from each of the secondary capitula.

γ. The structure of these threads; each consists of a single row of cells, containing in unripe specimens nucleated protoplasm; in older specimens each contains a coiled-up *antherozooid*.

b. *The antherozooids.*

α. Their form and structure; thickened at one end and granular; tapering off gradually towards the other end, which is hyaline and has two long cilia attached to it.

β. The movements in water of ripe antherozooids.

[Sometimes Chara cannot be obtained, when Nitella, another genus of the same Natural Order, and of similar habit and structure, can. Nearly all the points above described for Chara can be made out in Nitella, with the following differences: the cortical cells of the stem and leaves are absent, and, in the commoner species, the plant is not hardened by calcareous deposit; the branches arise, not *one* from a whorl of leaves, but *two;* and the five twisted cells of the spore-fruit are each capped by two small cells, instead of one.]

C. PROTOPLASMIC MOVEMENTS IN VEGETABLE CELLS.

a. **Chara.** Take a vigorous-looking fresh Chara or Nitella cell (say the terminal cell of a leaf), and examine it in water with a high power. Note

the superficial layer of protoplasm in which the
chlorophyll lies; it is stationary: focus through
this layer and examine the deeper one; note
the currents in it, marked by the granules they
carry along: their *direction;* in the long axis
of the cell, up one side and down the other,
the boundary of the two currents being marked
by the colourless band, in which no movements
occur. Try to find the nucleus; it has usually
disappeared in cells in which currents have
commenced, but when present is passive and
carried along by them. Sometimes it is very
difficult, on account of the incrustation of the
leaf-cells of Chara, to make out the protoplasmic
movements in them; if this is found to be the
case, the manubrial cells from an antheridium
should be used instead.

b.   **Tradescantia.** Examine in water, with a high
power, the hairs which grow upon the stamens :
they consist of a row of large roundish cells,
each with sac, protoplasm, nucleus, and vacuolar
spaces. Note the protoplasm; partly forming
a layer (*primordial utricle*) lining the sac and
heaped up round the nucleus, and partly forming bridles running across the cell in various
directions from the neighbourhood of the nucleus, and from one part of the protoplasm to
another; observe the currents in these bridles;
from the nucleus in some, towards it in others.

c.   **Vallisneria.** Take a leaf beginning to look old;
split it into two layers with a sharp knife and
mount a bit in water; examine with a high

power. Note the larger rectangular cells, belonging to the deeper layers, with well-marked currents in them, which carry the chlorophyll granules round and round inside the cell-wall.

If no currents are seen at first, gently warm the leaf by immersing it for a short time in water heated to a temperature between 30° and 35° C.

d. **Anacharis.** Take a yellowish-looking leaf: mount in water and examine with a high power; the phenomena observed are like those in Vallisneria. They are best observed in the single layer of cells at the margin of the leaf.

e. **Nettle-hair.** Mount an uninjured hair in water with the bit of leaf to which it is attached (it is essential that the terminal recurved part of the large cell forming the hair be not broken off); examine with the highest available power: currents carrying along very fine granules will be seen in the cell, their general direction being that of its long axis.

VII.

THE BRACKEN FERN (*Pteris aquilina*).

THE conspicuous parts of this plant are the large green leaves, or *fronds*, which rise above the ground, sometimes to the height of five or six feet, and consist of a stem-like axis or *rachis*, from which transversely disposed offshoots proceed, these ultimately subdividing into flattened leaflets, the *pinnules*. The rachis of each frond may be followed for some distance into the ground. Its imbedded portion acquires a brown colour, and eventually passes into an irregularly branched body, also of a dark-brown colour, which is commonly called the root of the fern, but is, in reality, a creeping underground stem, or *rhizome*. From the surface of this, numerous filamentous true roots are given off. Traced in one direction from the attachment of the frond, the rhizome exhibits the withered bases of fronds, developed in former years, which have died down; while, in the opposite direction, it ends, sooner or later, by a rounded extremity beset with numerous fine hairs, which is the apex, or growing extremity, of the stem. Between the free end and the fully formed frond one or more processes, the rudiments of fronds, which will attain their full development in following years, are usually found.

The attachments of the fronds are nodes, the spaces between two such successive attachments, internodes. It will be observed that the internodes do not become crowded towards the free end, and there is nothing comparable to the terminal bud of Chara with its numerous rudimentary appendages.

When the fronds have attained their full size, the edges of the pinnules will be observed to be turned in towards the underside, and to be fringed with numerous hair-like processes which roof over the groove, enclosed by the incurved edge.   At the bottom of the groove, brown granular bodies are aggregated, so as to form a streak along each side of the pinnule.   The granules are the *sporangia*, and the streaks formed by their aggregation, the *sori*.

Examined with a magnifying glass, each *sporangium* is seen to be pouch-shaped, like two watch-glasses united by a thick rim.   When ripe, it has a brown colour, readily bursts, and gives exit to a number of minute bodies which are the *spores*.

The plant now described is made up of a multitude of cells, having the same morphological value as those of *Chara*, and each consisting of a protoplasmic mass, a nucleus and a cellulose wall.   These cells, however, become very much modified in form and structure in different regions of the body of the plant, and give rise to groups of structures called *tissues*, in each of which the cells have undergone special modifications.   These tissues are, to a certain extent, recognizable by the naked eye.   Thus, a transverse section of the rhizome shews a *circumferential zone* of the same dark-brown colour as the external *epidermis*, enclosing a white *ground-substance*, interrupted by variously disposed *bands, patches*, and *dots*, some of which are of the same dark-brown hue as the external zone, while others are of a pale yellowish-brown.

The dark-brown dots are scattered irregularly, but the major part of the dark-brown colour is gathered into two narrow bands, which lie midway between the centre and the circumference. Sometimes the ends of these bands are united. Enclosed between these narrow, dark-brown bands are, usually, two elongated, oval, yellowish-brown bands; and, outside them, lie a number of similarly coloured patches, one of which is usually considerably longer than the others.

A longitudinal section shews that each of these patches of colour answers to the transverse section of a band of similar substance, which extends throughout the whole length of the stem; sometimes remaining distinct, sometimes giving off branches which run into adjacent bands, and sometimes uniting altogether with them.

At a short distance below the apex of the stem, however, the colour of all the bands fades away, and they are traceable into mere streaks, which finally disappear altogether in the semi-transparent gelatinous substance which forms the growing end of the stem. Submitted to microscopic examination, the white ground-substance, or *parenchyma*, is seen to consist of large polygonal *cells*, containing numerous starch granules; and the circumferential zone is formed of somewhat elongated cells, the thick walls of which have acquired a dark-brown colour, and contain little or no starch. The dark-brown bands, on the other hand, consist of cells which are so much elongated as almost to deserve the name of *fibres* and constitute what is termed *sclerenchyma*. Their walls are very thick, and of a deep-brown colour; but the thickening has taken place unequally, so as to leave short, obliquely directed, thin places, which look like clefts. The yellow bands, lastly, are *vascular bundles*. Each consists, externally, of thick-

walled, elongated, parallel-sided cells, internal to which lie
elongated tubes devoid of protoplasm, and frequently containing air. In the majority of these tubes, and in all
the widest, the walls are greatly thickened, the thickening
having taken place along equidistant transverse lines. The
tubes have become flattened against one another, by mutual
pressure, so that they are five- or six-sided; and, as the
markings of their flattened walls simulate the rounds of a
ladder, they have been termed *scalariform ducts* or *vessels.*
The cavities of these scalariform ducts are divided at
intervals, in correspondence with the lengths of the cells
of which they are made up, by oblique, often perforated,
partitions. Among the smaller vessels, a few will be found,
in which the thickening forms a closely wound spiral.
These are *spiral vessels.*

The rachis of a frond, so far as it projects above the
surface of the ground, is of a bright green colour; and, in
transverse section, it presents a green ground-substance,
interrupted by irregular paler markings, which are the transverse sections of longitudinal bands of a similar colour.
There are no brown spots or bands. Examined microscopically, the ground-substance is found to be composed
of polygonal cells containing chlorophyll. These are
invested superficially by an *epidermis*, composed of elongated cells, with walls thickened in such a manner as to
leave thin circular spots here and there. Hence, those
walls of the cells, which are at right angles to the axis of
vision, appear dotted with clear spots; while, in those
walls of which transverse sections are visible, the dots are
seen to be funnel-shaped depressions.

The pale bands are vascular bundles containing scalariform and spiral vessels. The outer layer investing each
is chiefly formed of long hollow fibres with very thick

·walls, and terminating in a point at each end. These sclerenchymatous *fibres* have oblique cleft-like clear spaces, produced by interruptions of the process of thickening in their walls.

The vascular bundles, the green parenchyma, and the epidermis are continued into each pinnule of the frond. The epidermis retains its ordinary character on the upper side of the pinnule, except that the contours of its component cells become somewhat more irregular. On the under side, many hairs are developed from it, and the cells become singularly modified in form, their walls being thrown out into lobes, which interlock with those of adjacent cells.

Between many of these cells an oval space is left, forming a channel of communication between the interior of the frond and the exterior. The opening of this space is surmounted by two reniform cells, the concavities of which are turned towards one another, while their ends are in contact. The opening left between the applied concave faces is a *stomate;* and, as the *stomata* are present in immense numbers, there is a free communication between the outer air and the *intercellular passages* which exist in the substance of the frond. Those cells of the green parenchyma of the frond which form the inferior half of its thickness, in fact, are irregularly elongated, and frequently produced into several processes, or stellate. ·They come into contact with adjacent cells only by comparatively small parts of their surfaces, or by the ends of these processes. They thus bound passages between the cells, *intercellular passages*, which are full of air, and are in communication with similar, but narrower, passages, which extend throughout the substance of the plant.

The vascular bundles break up in the pinnules, and

follow the course of the so-called *veins* which are visible upon its surface; ducts being continued into their ultimate ramifications.

The rootlets present an outer coat of epidermis, enclosing parenchyma traversed by a central vascular bundle. They increase in length by the division and subdivision of the cells at the growing point, but this point is not situated at the very surface of the rootlet, as the growing point at the extremity of the rhizome is, but is covered by a cap of cells.

When the spores are sown upon damp earth, or a tile, or a slip of glass, and kept thoroughly moist and warm, they germinate. Each gives rise to a tubular, hypha-like prolongation, which developes a similar process, the *primitive rootlet*, close to the spore. The hypha-like prolongation, at first, undergoes transverse division, so that it becomes converted into a series of cells. Then, the cells at its free end divide longitudinally, as well as transversely, and thus give rise to a flat expansion, which gradually assumes a bilobed form, and becomes thickened, in some parts, by division of its cells in a direction perpendicular to its surface. The protoplasm of these cells developes chlorophyll granules, whereby the bilobed disk acquires a green colour; while numerous simple radicle fibres are given off from its under surface, and attach the little plant, which is termed a *prothallus* or *prothallium*, to the surface on which it grows.

The prothallus attains no higher development than this, and does not directly grow into a fern such as that in which the spores took their origin; but, after a time, rounded or ovoidal elevations are developed, by the outgrowth and division of the cells which form its under aspect. Some of these are *antheridia*. The protoplasm of each of the cells contained in their interior is converted into an antherozooid, somewhat similar to that of *Chara*, but provided with many more cilia. The antheridium bursts, and the antherozooids, set free from their containing cells, are propelled through the moisture on the under surface of the prothallus by their cilia.

The processes of the second kind acquire a more cylindrical form, and are called *archegonia*. Of the cells which are situated in the axis of the cylinder, all disappear but that which lies at the bottom of its cavity. This is the *embryo cell*, and when the archegonium is fully formed, a canal leads from its summit to this cell. The antherozooids enter by this canal, and impregnate the embryo cell.

The embryo cell now begins to divide, and becomes converted into four cells; of these, the two which lie at the deepest part of the cavity of the archegonium subdivide and ultimately form a plug-like, cellular, mass, which imbeds itself firmly in the substance of the prothallus. Of the remaining two cells, which also undergo subdivision, one gives rise to the rhizome of the young fern, while the other becomes its first rootlet. It appears probable that the plug-like mass absorbs nutritive matter from the prothallus, and supplies the rhizome of the young fern, until it is able to provide for itself. As the rhizome grows, and developes its fronds, it rapidly attains a size vastly superior to that of the prothallus, which at length ceases to have any functional importance, and disappears.

Thus *Pteris* presents a remarkable case of the alternation of generations. The large and complicated organism commonly known as the 'Fern' is the product of the impregnation of the embryo cell by the antherozooid. This 'Fern,' when it attains its adult condition, developes sporangia; and the inner cells of these sporangia give rise, by a perfectly asexual fissive process, to the spores. The spores when set

free germinate; the product of that germination is the incon-
spicuous and simply cellular prothallus; an independent
organism, which nourishes itself and grows, and on which,
eventually, the essential organs of the sexual process—the
archegonia and antheridia—are developed.

Each impregnated embryo cell produces only a single
'fern,' but each 'fern' may give rise to innumerable prothallia, seeing that every one of the numerous spores developed in the immense multitude of sporangia to which the
frond gives rise, may germinate.

LABORATORY WORK.

A.   THE FERN-PLANT; ASEXUAL GENERATION.

a. External characters.

   a.   The brown underground stem or *rhizome*, with
      a lighter band (the *lateral line*) running along
      each side of it: its *nodes* and *internodes*.

   b.   The *roots* springing from the rhizome.

   c.   The *leaves* or *fronds* arising from the rhizome at
      intervals, along the lateral lines.

      a.   The great amount of subdivision of the frond:
         its main axis (*rachis*); the primary divisions or
         pinnæ; the ultimate divisions or *pinnules.*

      β.   The *sori;* small brown patches along the
         margin of the under surface of some of the
         pinnules.

   d.   The *nodes* and *internodes* of the rhizome.   The
      absence of a terminal bud on it.

b. The rhizome.

1. Cut it across and draw the section as seen with the naked eye.

    a. The outer brownish layer (*epidermis* and *sub-epidermis*); the latter thins away somewhat, opposite the lateral lines.

    b. The yellowish-white substance (*ground-substance* or *parenchyma*) forming most of the thickness of the section.

    c. The internal incomplete brown ring (*sclerenchyma*) imbedded in the parenchyma.

    d. The small patches of sclerenchyma scattered about in the parenchyma outside the main sclerenchymatous ring.

    e. The yellowish tissue (*vascular bundles*) lying inside and outside the ring of sclerenchyma.

2. Cut a longitudinal section of the rhizome; make out on the cut surface **b.** 1. *a, b, c, d.*

3. Cut a thin transverse section of the rhizome, mount in water and examine with 1 inch obj.

    a. The single layer of much thickened epidermic cells.

    b. The small opaque angular contours of the sub-epidermic cells (*external sclerenchyma*).

    c. The large polyhedral more transparent parenchymatous cells.

    d. The small opaque angular contours of the cells of the internal sclerenchyma.

    e. The great openings of the ducts and vessels in the fibro-vascular bundles.

    Draw the section.

4.   Examine with $\frac{1}{8}$ obj.

 a.   The *epidermis* : its thick-walled cells:

 b.   The *parenchyma* : its large thin-walled cells : their
      sac, protoplasm and nucleus : the great number
      of starch granules in them.

 c.   The various patches of *sclerenchyma*, made up of
      thick-walled angular cells.

 d.   The *vascular bundles*.   Note in each :—

   a.   Outside, a single layer of cells containing no
        starch granules (*bundle sheath*).   These really
        belong to the parenchyma or ground tissue.

   β.   Within the bundle sheath a layer of small
        parenchymatous cells containing starch (*inner*
        or *bast sheath*).

   γ.   Within the last layer comes the *bast* of the
        bundle (*phloem*) consisting of—externally, two
        or more layers of small rectangular cells with
        thickened walls (*bast fibres*) and then a single
        row of large thin-walled cells (*bast vessels*)
        between which lie smaller thin-walled cells
        containing starch granules (*bast parenchyma*).

   δ.   Within the bast are seen the cross sections
        of the *vessels:* note their greatly thickened
        walls, and large central cavity containing no
        protoplasm.

   ε.   Scattered here and there, in the spaces between
        the angles of the vessels, are small parenchy-
        matous cells (*wood parenchyma*) containing
        starch granules.

        The wood, or xylem, consists of δ and ε.

   ζ.   Treat with iodine : the protoplasm stained
        brown; the starch granules deep blue, render-

ing some of the cells quite opaque and almost black-looking.

5.   Cut a thin longitudinal section of the stem and examine with 1 inch and then with $\frac{1}{8}$ obj.   Make out the various tissues described in 3 and 4.

a.   The *epidermis*, subepidermis and *parenchyma*, much as in the transverse section, except that the subepidermic cells are longer.

b.   The *sclerenchyma* is seen to be made up of greatly elongated cells, tapering towards each end.

c.   The vascular bundles; note in them—

α.   The cells of the bundle sheath much as in the transverse section; the *bast fibres*, elongated, with thickened walls; the cells of the bast parenchyma somewhat elongated ; the *bast vessels*, elongated cells, presenting irregular patches of pores (*sieve-tubes*); the *bast sheath* cells somewhat elongated.

β.   The *vessels:* elongated tubes presenting cross partitions, dividing them into separate cells, at long intervals.   Two forms of vessel will be seen, viz. *scalariform vessels*, with regular transverse thickenings on their walls and *spiral vessels*, less numerous than the last form: with a continuous spiral thickening on their walls.

γ.   The *bast cells:* seven or eight times as long as they are broad, and terminating obliquely at each end.

δ.   The elongated larger cells (4. *d.* δ): they have very slightly thickened walls and no scalariform markings.

6. [Cut off half an inch of the growing end of the stem,
imbed it in paraffin upside down, and cut a series of
transverse sections: examine them with the microscope,
beginning with those farthest from the growing point.
At first the various tissues described in 3 and 4 will be
readily recognisable; as the sections nearer the growing point are examined they will be less distinct, and
close to the growing point the whole section will be
found to be composed entirely of parenchymatous
closely-fitting cells.]

[c. **The leaf.** Imbed a leaf in paraffin and cut a thin
vertical section: examine with 1 inch obj. It will be
found to be constructed essentially on the same plan
as the leaf of the bean. · (VIII.)]

d. **The reproductive organs.**

1. Examine a *sorus* with a low power without a cover-
glass. It is composed of a great number of minute
oval bodies, the *sporangia.*

2. Scrape off some sporangia and mount in water: examine with 1 inch obj.

   a. *Their form:* they are oval biconvex bodies
   borne on a short stalk.

   b. *Their structure:* composed of brownish cells, one
   row of which has very thick walls, and forms a
   marked ring (*annulus*) round the edge of the
   sporange.

   c. Their mode of *dehiscence* (look out for one that
   has opened): by a cleft running towards the
   centre of the sporange from a point where the
   annulus has torn across.

3. Burst open some sporangia by pressing on the cover-
glass: examine, with $\frac{1}{8}$ obj., the *spores* which are set
free.

a.   *Their size:* measure.

b.   *Their form:* somewhat triangular.

[c.   *Their structure:* a thick outer coat, a thin inner coat, protoplasm, and a nucleus: crush some by pressure on the cover-glass.]

B.   THE PROTHALLUS; SEXUAL GENERATION.

Prothalli may be obtained by sowing some spores on a glass slide, and keeping them warm and very moist for about three months.   They are small deep green leaf-like bodies.

a.   **The Prothallus.**

1.   Transfer a prothallus to a slide, and mount it in water with its under surface uppermost.   Examine with 1 inch obj.

a.   *Its form:* a thin kidney-shaped expansion from which, especially towards its convex border, a number of slender filaments (*rootlets*) arise.

b.   *Its structure.*

α.   *The leafy expansion :* it consists throughout most of its extent of a single layer of polyhedral chlorophyll-containing cells, but at a part (the *cushion*) a little behind the depression marking the growing point it is several cells thick.

β.   *The rootlets:* composed of a series of cells which contain no chlorophyll.

c.   The *antheridia* and *archegonia:* the former can just be seen with an inch objective as minute eminences on the under surface of those parts of the prothallus which consist of a single layer of cells, especially among the root-hairs ; the latter are partly imbedded in the cushion.

b. The reproductive organs.

These are to be found by examining the under surface of the prothallus with ⅛ obj.

1. The *antheridia.* Most numerous near and among the rootlets.

   a. *Their form:* small hemispherical eminences.

   b. *Their structure:* made up of an outer layer of cells containing a few chlorophyll-granules, and through which can be seen, according to the stage of development, either a single *central cell,* or a number of smaller cells (*mother-cells of antherozooids*) resulting from its division: in the latter cells, in ripe antheridia, spirally coiled bodies (*antherozooids*) can be indistinctly seen.

2. *The antherozooids.*

   Some of these are sure to be found swimming about in the water if a number of ripe prothalli are examined.

   a. Small bodies, coiled like a corkscrew, thick at one end, and tapering towards the other, which has a number of cilia attached to it. To the thicker end of the antherozooid is often attached a rounded mass containing colourless granules.

   b. Treat with iodine; this stains them and stops their movements, so that their form can be more distinctly seen.

3. *The archegonia.* Make vertical sections of the prothallus passing through the cushion; either by simply chopping down it with a razor, or holding it between two pieces of carrot and cutting. Note in the archegonia—

   a. *Their form:* chimney-shaped eminences with a small aperture at the apex.

b.   *Their structure.*   Each is composed of a layer of transparent cells containing no chlorophyll, arranged in four rows, and surrounding a central cavity which extends into the cushion formed by the thickened part of the prothallus (a. 1. b. a). In this cavity lies, in young specimens, a large nucleated granular basal cell, with two or three smaller granular cells (*neck-cells*) above it in the narrow upper part of the cavity; in older specimens this upper part is empty, forming a canal leading down to the basal cell.

4.   Examine young Fern in connection with its prothallus.

THE BEAN-PLANT (*Vicia faba*).

IN this, which is selected as a convenient example of a Flowering Plant, the same parts are to be distinguished as in the Fern; but the axis is erect and consists of a *root* imbedded in the earth and a *stem* which rises into the air. The appendages of the stem are *leaves*, developed from the opposite sides of successive nodes; and the internodes become shorter and shorter towards the summit of the stem, which ends in a terminal *bud.* Buds are also developed in the axils of the leaves, and some of them grow into branches, which repeat the characters of the stem; but others, when the plant attains its full development, grow into stalks which support the *flowers;* each of which consists of a *calyx*, a *corolla*, a *staminal tube* and a central *pistil;* the latter is terminated by a *style*, the free end of which is the *stigma*.

The staminal tube ends in ten filaments, four of which are rather shorter than the rest; and the filaments bear oval bodies, the *anthers*, which, when ripe, give exit to a fine powder, made up of minute *pollen* grains. The pistil is hollow; and, attached by short stalks along the ventral side of it, or that turned towards the axis, is a longitudinal series of minute bodies, the *ovules*. Each ovule consists of a central conical *nucleus*, invested by two coats, an *outer* and an *inner*. Opposite the summit of the nucleus, these coats are perforated by a canal, the *micropyle*, which leads down to the

nucleus. The nucleus contains a sac, the *embryo sac*, in
which certain cells, one of which is the *embryo cell*, and
the rest *endosperm* cells, are developed. A pollen grain
deposited on the stigma, sends out a hypha-like prolongation, the *pollen tube*, which elongates, passes down the style,
and eventually reaches the micropyle of an ovule. Traversing the micropyle, the end of the pollen tube penetrates the
nucleus, and comes into close contact with the embryo sac.
This is the process of impregnation, and the result of it
is that the embryo cell divides and give rise to a cellular
embryo. This becomes a minute Bean-plant, consisting of a
radicle or primary root; of two, relatively large, primary
leaves, the *cotyledons;* and of a short stem, the *plumule*, on
which rudimentary leaves soon appear. The cotyledons now
increase in size, out of all proportion to the rest of the em-
bryonic plant; and the cells of which they are composed
become filled with starch and other nutritious matter. The
nucleus and coats of the ovule grow to accommodate the
enlarging embryo, but, at the same time, become merged
into an envelope which constitutes the coat of the seed. The
pistil enlarges and becomes the pod; this, when it has
attained its full size, dries and readily bursts along its edges,
or decays, setting the seeds free. Each seed, when placed
in proper conditions of warmth and moisture, then germinates.
The cotyledons of the contained embryo swell, burst the
seed coat, and, becoming green, emerge as the fleshy *seed
leaves.* The nutritious matters which they contain are absorbed by the plumule and radicle, the latter of which descends into the earth and becomes the root, while the former
ascends and becomes the stem of the young bean-plant.
The apex of the stem retains, throughout life, the simply
cellular structure which is, at first, characteristic of the whole
embryo; and the growth in length of the stem, so far as it

depends on the addition of new cells, takes place chiefly, if not exclusively, in this part.

The apex of the root, on the other hand, gives rise to a root-sheath, as in the Fern.

The leaves cease to grow by cell multiplication at their apices, when these are once formed, the addition of new cells taking place at their bases.

The tissues which compose the body of the Bean-plant are similar, in their general characters, to those found in the Fern, but they differ in the manner of their arrangement. The surface is bounded by a layer of epidermic cells, within which, rounded or polygonal cells make up the ground-substance, or parenchyma, of the plant, extending to its very centre in the younger parts of the stem and in the root; while, in the older parts of the stem, the centre is occupied by a more or less considerable cavity, full of air. This cavity results from the central parenchyma becoming torn asunder, after it has ceased to grow, by the enlargement of the peripheral parts of the stem. Nearer to the circumference than to the centre, lies a ring of woody and vascular tissue, which, in transverse sections, is seen to be broken up into wedge-shaped bundles, by narrow bands of parenchymatous tissue, which extend from the parenchyma within the circle of woody and vascular tissue (*medulla* or pith) to that which lies outside it. Moreover, each bundle of woody and vascular tissue is divided into two parts, an outer and an inner, by a thin layer of small and very thin-walled cells, termed the *cambium* layer. What lies outside this layer belongs to the *bark* and *epidermis;* what lies inside it, to the *wood* and *pith.*

The great morphological distinction between the axis of the Bean and that of the Fern lies in the presence of this cambium layer. The cells composing it, in fact, retain

their power of multiplication, and divide by septa parallel
with the length of the stem, or root, as well as transverse to
it. Thus new cells are continually being added, on the
inner side of the cambium layer, to the thickness of the
wood, and on the outer side of it, to the thickness of the
bark; and the axis of the plant continually increases in
diameter, so long as this process goes on. Plants in which
this constant addition to the outer face of the wood and the
inner face of the bark takes place, are termed *exogens.*

At the apex of the stem, and at that of the root, the
cambium layer is continuous with the cells, which retain
the capacity of dividing in these localities. As the plant is
thickest at the junction of the stem and root, and diminishes
thence to the free ends, or apices, of these two structures,
the cambium layer may be said to have the form of a double
cone. And it is the special peculiarity of an exogen to
possess this doubly conical layer of constantly dividing
cells, the upper end of which is free, at the growing point
of the terminal bud of the stem, while its lower end is
covered by the root-cap of the ultimate termination of the
principal root.

The most characteristic tissues of the wood are dotted
ducts and spiral vessels, the spiral vessels being particularly
abundant close to the pith. The bark contains elongated
liber or bast cells; but there are no scalariform vessels such
as are found in the Fern.

Stomates are absent in the epidermis of the root: they
are to be found, here and there, in the epidermis of all the
green parts of the stem and its appendages, but, as in the
Fern, they are most abundant in the epidermis of the under
side of the leaves. As in the Fern, they communicate with
intercellular passages, which are widest in the leaves, but
extend thence throughout the whole plant.

The difference between a flowering plant, such as the Bean, and a flowerless plant, such as the Fern, at first sight appears very striking, but it has been proved that the two are but the extreme terms of one series of modifications. The *anther*, for example, is strictly comparable to a *sporangium*. The *pollen grains* answer to the male *spores* of those flowerless plants in which the spores are of distinct sexes— some spores giving rise to prothallia which develope only antheridia, and others to prothallia which develope only archegonia; instead of the same prothallia producing the organs of both sexes, as in *Pteris*. And the *pollen tube* corresponds with the first *hypha-like process* of the spore. But, in the flowering plants, the protoplasm of the pollen tube does not undergo division and conversion into a prothallus, from which antheridia are developed, giving rise to detached fertilizing bodies or antherozooids, but exerts its fertilizing influence without any such previous differentiation. The connecting links between these two extreme modifications are furnished, on the one hand, by the Conifers, in which the protoplasm of the pollen tube becomes divided into cells, from which, however, no antherozooids are developed; and the Club-mosses, in which the protoplasm of the male spores (= pollen grains) divides into cells which form no prothallus, but give rise directly to antherozooids.

On the other hand, the *embryo sac* is the equivalent of a female *spore:* the *endosperm* cells, which are produced from part of its protoplasm, answer to the cells of a *prothallus;* while the *embryo cell* of the flowering plant corresponds with the *embryo cell* contained in the archegonium of the prothallus. In the development of the female spore of the flowering plant, therefore, the free prothallus and the archegonia are suppressed. Here, again, the intermediate stages are presented by the Conifers and the Club-mosses. For, in the Conifers, the protoplasm of the embryo sac gives rise to a solid prothallus-like endosperm, in which bodies called *corpuscula*, which answer to the archegonia, are formed; and in these the embryo cells arise; while, in some of the Club-mosses, there are female spores distinct from the male spores, and the prothallus which they develope does not leave the cavity of the spore, but remains in it like an endosperm.

The physiological processes which go on in the higher green plants, such as the Fern and the Bean, resemble, in the gross, those which take place in *Protococcus* and *Chara*. For such plants grow and flourish if their roots are immersed in water containing a due proportion of certain saline matters, while their stem and leaves are exposed to the air, and receive the influence of the sun's rays.

A Bean-plant, for instance, may be grown, if supplied through its roots with a dilute watery solution of potassium and calcium nitrate, potassium and iron sulphate, and magnesium sulphate. While growing it absorbs the solution, the greater part of the water of which evaporates from the extensive surface of the plant. In sunshine, it rapidly decomposes carbonic anhydride, fixing the carbon, and setting free the oxygen; at night, it slowly absorbs oxygen, and gives off carbonic acid; and it manufactures a large quantity of protein compounds, cellulose, starch, sugar and the like, from the raw materials supplied to it.

It is further clear that, as the decomposition of carbonic anhydride can take place only under the combined influences of chlorophyll and sunlight, that operation must be confined, in all ordinary plants, to the tissue immediately beneath the epidermis in the stem, and to the leaves. And it can be proved, experimentally, that fresh green leaves possess this power to a remarkable extent.

On the other hand, it is clear that, when a plant is grown under the conditions described, the nitrogenous and mineral constituents of its food can reach the leaves only by passing from the roots, where they are absorbed, through the stem to the leaves. And, at whatever parts of the plant the nitrogenous and mineral constituents derived from the roots are combined with the carbon fixed in the leaves, the resulting compound must be diffused thence, in order to reach the deep-seated cells, such for instance as those of the cambium layer and those of the roots, which are growing and multiplying, and yet have no power of extracting carbon directly from carbonic anhydride. In fact, those cells which contain no chlorophyll, and are out of the reach of light, must live after the fashion of *Torula;* and manufacture their protein out of a material which contains nitrogen and hydrogen, with oxygen and carbon, in some other shape than that of carbonic anhydride. The analogy of *Torula* suggests a fluid which contains in solution, either some ammoniacal salt comparable to ammonium tartrate, or a more complex compound analogous to pepsin. Thus, the higher plant combines within itself the two, physiologically distinct, lower types of the Fungus and the Alga.

That some sort of circulation of fluids must take place in the body of a plant, therefore, appears to be certain, but the details of the process are by no means clear. There is evidence to shew that the ascent of fluid from the root to the leaves takes place, to a great extent, through the elongated ducts of the wood, which not unfrequently open into one another by their applied ends, and, in that way, form very fine capillary tubes of considerable length.

The mechanism by which this ascent is effected is of two kinds; there is a pull from above, and there is a push from below. The pull from above is the evaporation which takes place at the surface of the plant, and especially in the air-passages of the leaves, where the thin-walled cells of the parenchyma are surrounded, on almost all sides, with air, which communicates directly with the atmosphere through the stomates. The push from below is the absorptive action which takes place at the extremities of the rootlets, and which, for example, in a vine, before its leaves have grown in the spring, causes a rapid ascent of fluid (*sap*) absorbed from the soil. A certain portion of the fluid thus pumped up from the roots to the surface of the plant doubtless exudes, laterally, through the walls of the vessels (the thin places which give rise to the dots on the walls of these structures especially favouring this process), and, passing from cell to cell, eventually reaches those which contain chlorophyll. The distribution of the compound containing nitrogen and carbon, whatever it may be, which is formed in the chlorophyll-bearing cells, probably takes place by slow diffusion from cell to cell.

The supply of air, containing carbonic anhydride, to the leaves and bark is effected by the abundant and large air-passages which exist between the cells in those regions. But it can hardly be doubted that all the living protoplasm of the plant undergoes slow oxidation, with evolution of carbonic anhydride; and that this process, alone, takes place in the deeper-seated cells. The supply of oxygen needful for this purpose is sufficiently provided for, on the one hand, by the minute air-passages which are to be found between the cells in all parenchymatous tissues; and on the other, by the spiral vessels, which appear always to contain air under normal circumstances, in the woody bundles. The replacement of the oxygen of the air thus absorbed, and the removal of the carbonic anhydride formed, will be sufficiently provided for by gaseous diffusion.

From what has been said, it results that, in an ordinary plant, growing in damp earth and exposed to the sunshine, a current of fluid is setting from the root towards the surface exposed to the air, where its watery part is for the most part evaporated; while gaseous diffusion takes place, in the contrary direction, from the surface exposed to the air, through the air-passages and. spiral vessels which extend from the stomates to the radicles; the balance of exchange being in favour of oxygen, in all the chlorophyll-bearing parts of the plant which are reached by the sunlight, and in favour of carbonic anhydride, in its colourless and hidden regions. At night, the evaporation diminishing with the lowering of the temperature, the ascent of liquid becomes very slow, or stops, and the balance of exchange in the air-passages is entirely in favour of carbonic anhydride; even the chlorophyll-bearing parts oxydizing, while no carbonic anhydride is decomposed.

LABORATORY WORK.

a.   **General characters.**

  a.   The erect central main axis (*root* and *stem*).

  b.   The *branches:* some, mere repetitions of the main axis; others, modified and bearing flowers.

  c.   The *nodes* and *internodes.*

  d.   The *appendages.*

   α.   Rootlets.

   β.   Foliage leaves.

   γ.   Floral leaves.

b.   **The root.**

a.   Its main central portion (*axis*).

b.   The irregularly arranged *rootlets* attached to the axis.

c.   The absence of chlorophyll in the root.

d.   The *root-sheath*, covering the tip of each rootlet: this is difficult to get whole out of the ground in the bean, but is readily seen by examining the roots of duckweed (*Lemna*) with 1 inch obj. In the latter plant it consists of several layers of cells forming a cap on the end of the root, and ending abruptly with a prominent rim some way up it.

c.   **The stem.**

1.   Erect, green, four-cornered, with a ridge at each angle; not woody; the gradual shortening of the internodes towards its apex.

2.   Cut a thin transverse section of the stem through an internode; note its central cavity, and the whitish ring of *fibro-vascular bundles* in it, which is harder to cut than the rest: mount in water and examine with 1 inch obj.: note—

a.   The *medullary* or *pith-cavity* in the centre of the section.

b.   The *pith-cells*, around the central cavity: large and more or less rounded (*parenchyma*): sometimes with dotted walls from spots of local thinness on them.

c.   The *epidermis:* composed of a single layer of somewhat squarish-looking cells, containing no chlorophyll.

d.   Beneath the epidermis several layers of large
      rounded cells containing chlorophyll (*parenchyma
      of the bark*).

e.   The *medullary rays:* radiating rows of paren-
      chymatous cells uniting *b* and *d:* not quite con-
      tinuous, being interrupted by the *cambium zone*
      (*f. γ*).

f.   The *fibro-vascular bundles*, lying between the
      medullary rays; commencing at the side nearest
      the pith, note—

    a.   The large openings formed by the transverse
            sections of the spiral *vessels* and *ducts.*

    β.   The small thick-walled *wood-cells*, wedged in
            between the vessels. These two (*a* and *β*)
            form the wood or xylem of the bundle. The
            bast or *phloem.* It presents internally thin-
            walled cells of various size, the bast paren-
            chyma and bast vessels or sieve tubes. Ex-
            ternally it appears in cross section to be com-
            posed of rounded cells with thickened walls;
            the *bast fibres* or *liber.*

    γ.   The *cambium zone:* granular-looking, and
            composed of small angular thin-walled cells.

    δ.   The *liber-layer:* in cross section it seems
            composed of rounded cells with much thick-
            ened walls. Draw the section.

3.   Cut a transverse section through a node, and compare
     it with that through the internode.

4.   Cut a thin longitudinal section through part of an
     internode (if necessary the bit of stem may be im-
     bedded in paraffin first), and mount it in water;

working from the medullary cavity outwards, note
the following layers, using at first a low power :—

a. *The pith-cells:* much as in the transverse section.

b. The fibro-vascular bundles presenting—

α. *The spiral vessels:* elongated tubes with a spiral
thickening on their walls.

β. *The wood-cells:* elongated and with much
thickened walls.

γ. *The dotted ducts:* much like *b*, but the thickening not deposited in the form of a spiral.

δ. *The cambium zone:* made up of cloudy-
looking, small, angular, thin-walled cells.

ε. The *bast parenchyma:* thin-walled elongated
cells.

ζ. The *bast vessels:* larger elongated cells with
oblique perforated septa (*sieve-tubes*).

η. The bast fibres, fusiform and thick-walled.

c. More parenchymatous cells.

d. *Epidermis:* composed apparently of cubical
colourless cells: here and there the opening of
a stomate (**d. 2.** *d.* β) may be seen.

Draw the section.

5. Compare the transverse and longitudinal sections
together, making out the corresponding parts in each.

6. Put on a high power, and examine each of the above-
mentioned tissues carefully.

7. Stain with iodine: note the *cell-walls;* the *protoplasm*
—its presence or absence, and relative quantity in
the various tissues; the *nuclei* of the cells; *starch
granules* in some, stained deep blue by the iodine.

d. The leaves.

1. *Their form and composition.*

a. Each leaf consists of a number of different parts, viz. :—

   a. The stalk or *petiole.*

   β. The four to six oval *leaflets* attached laterally to the stalk.

   γ. The pair of small leaf-like expansions (*stipules*) at the base of the petiole.

   δ. The rudimentary *tendril* terminating the petiole.

2. *The histological structure of a leaflet.*

a. Imbed a leaflet in paraffin or hold it between two bits of carrot or turnip and cut a thin section from it, perpendicular to its surfaces. Let the section lie in water a few minutes to drive the air out of its intercellular spaces, and then mount it in water, and examine with 1 inch objective.

b. Begin at the upper surface (marked out by its more closely packed cells), and work through to the lower. Note—

   a. The colourless *epidermic layer*—consisting of a single row of cells; the openings here and there in it (*stomata*).

   β. Beneath the upper epidermis come elongated chlorophyll-containing cells, set on perpendicularly to the surface.

   γ. Then come irregularly branched (*stellate*) cells forming the lower half of the leaf-substance; these also contain chlorophyll.

δ. The epidermic layer of the lower surface; like α.

ε, The *intercellular spaces*, through the whole thickness of the leaf: the direct communication of some of them with stomata.

ζ. Here and there sections of *ribs* or *veins:* make out in them the same elements as in c. 2. *f.*

Draw.

c. Treat with iodine: make out the sac, protoplasm (*primordial utricle*), nucleus and vacuole of the cells: the starch granules.

d. Peel off a strip of epidermis from a leaf and examine with a low power: note—

   α. The large close-fitting cells, with irregularly wavy margins and no chlorophyll, which chiefly make up the epidermis.

   β. The openings here and there in it (*stomata*); the two curved, chlorophyll-containing cells bounding each stomate.

   e. Gently pull a midrib in two across its long axis; note the fine threads uniting the two broken ends; cut them off with a sharp pair of scissors, mount in water and examine with $\frac{1}{4}$ or $\frac{1}{8}$ objective: they will be found to consist of partially unrolled spiral vessels.

e. The flower.

1. *Its general structure.*

   a. Borne on a short stalk (*peduncle*).

   b. Composed of four rows or *whorls* of organs.

      α. The external green cup-like *calyx.*

β. Inside the calyx the *corolla :* the most conspicuous part of the flower.

γ. Inside the corolla the *stamens*.

δ. Within the stamens the *pistil.*

2. *The calyx.*

A cup terminated at its free edge by five prominent
points, two dorsal, and three ventral : the five small
midribs running along it (one to the end of each of
the points) represent the free ends of five *sepals,*
which are united below.

3. *The corolla.*

a. Composed of five pieces or *petals.*

α. On the dorsal side, a single large piece (*vexil-
lum*) expanded at its free end and folded over
the rest.

β. On the sides, two oval pieces (*alæ*), each
attached by a distinct narrowed stalk (*unguis*).

γ. The inferior part of the corolla (*carina*), composed of two oval pieces united along their
lower edge but readily tearing apart.

4. *The stamens.*

a. Ten in number, each consisting of a stalk-like
part, *the filament,* terminated by a small knob,
the anther.

b. The union of the filaments for three-fourths of
their length to form the *stamen-tube :* the sharp
bend of the filaments towards the upper side at
the point where they separate from one another.

c. Tease out an anther in water and examine with
$\frac{1}{8}$ obj. : there will be found numerous—

a. *Pollen-grains:* small oval bodies, with projections on them in the equatorial region.

d. The anther of a bean is so small that sections cannot be made of it without considerable skill: the structure of an anther can however be easily made out by imbedding one from a tiger-lily in paraffin or holding it between two bits of carrot, cutting transverse sections, mounting in water and examining with 1 inch obj.

a. It contains four chambers, two on each side of the continuation of the filament, and in each chamber lie numerous pollen-grains.

5. *The pistil.*

a. It is found by tearing open the stamen-tube: it is a long green tapering body, somewhat flattened laterally and ending in a point (the *style*) which bears a' tuft of strong hairs.

b. Slit it open carefully: in it lies a central cavity, containing a number of small oval bodies, *the ovules*, attached along its ventral side by short pedicles.

c. It is difficult to get a section of a bean-ovule, but its essential structure may be readily made out by making thin transverse sections of the ovary of a large lily (where the ovules are closely imbedded in a large quantity of parenchyma) and examining with 1 inch obj.

a. The central cellular portion of the ovule (*nucleus*) made up of a large number of cells.

β. Its two coats, an inner (*primine*) and outer (*secundine*).

γ.   The small passage (*micropyle*) leading through the coats down to the nucleus.

δ.   In some specimens, a large cavity (the *embryo-sac*) will be seen in the nucleus just opposite the micropyle. In the embryo-sac may be seen some small granular cells (the *embryo-cell* and *endosperm cells*).

f. The seeds.

1. Soak some dried beans in water for twenty-four hours; they will slightly swell up and be more readily examined than when dry.

    a.   Note the black patch on one end of the bean, marking where the stalk (*funiculus*) which fixed it in the pod was attached to it.

    b.   Having wiped all moisture off the bean gently press it while observing that part of the black patch which is next its broader end : close to the patch a minute drop of fluid will be observed to be pressed out through a small opening, the *micropyle*.

    c.   Carefully peel off the outer coat (*testa*) of the seed : the two large fleshy *cotyledons* will be laid bare.

    d.   Joining the cotyledons together will be found the rest of the embryo : it consists of a conical part (the *radicle*) lying outside the cotyledons, with its apex directed towards the point where the micropyle was; and of the rudiments of the stem and leaves (*plumule*) lying between the cotyledons.

g. The process of fertilization.

This is difficult to follow in the bean; but by using different plants for the observation of its various stages it is fairly easy to observe all its more important steps.

1. A plant well adapted for seeing the penetration of the pollen-tube into the stigma and style is the Evening Primrose (*Œnothera biennis*).

   Detach the style from the flower and hold the club-shaped stigma between the finger and thumb of the left hand. Moisten it with a drop of water and then make with a wetted razor several successive cuts through it. This will divide the stigma into several slices. Spread these out on a glass slide with a needle in water and examine the thinnest, after putting on a covering-glass.

   The triangular grains of pollen will be seen sending out from one angle a tube into the stigmatic tissue, which is easily seen from its slight difference in colour.

2. The entrance of the pollen-tube into the micropyle can be readily made out in some species of *Veronica.* The common *V. serpyllifolia* — often to be found in shady places on lawns—is well adapted for the purpose. A flower should be taken from which the corolla has *just* dropped. Dissect out the minute ovary and, using the dissecting microscope, open with a needle one of its two cells in a drop of water; remove the mass of ovules and gently tease them apart. Then put on a covering-glass and examine with a low power till an ovule is found which shews the entry of the pollen-tube. The addition of dilute glycerine will make the ovule more transparent, so

that after some time the embryo-sac can be seen,
and the progress of the pollen-tube into the ovule
followed.

3. The young fruit of *Campanula* (especially the common Canterbury Bells of gardens, *Campanula me-
dia*) is convenient for examining the embryo-sac.
It is only necessary to cut thin transverse sections
of the fruit and examine in water. Some of the
ovules cut through will allow the embryo-sac to be
seen, and in fortunate sections the embryo-vesicle
and the end of the pollen-tube in contact with the
embryo-sac.

THE BELL-ANIMALCULE (*Vorticella*).

THE great majority of those animal organisms which are more complex than *Amœba*, begin their existence as simple nucleated cells, having a general similarity to *Amœba;* and the single nucleated cell which constitutes the whole animal in its primitive condition divides and subdivides until an aggregation of similar cells is formed. And it is by the differentiation and metamorphosis of these primitively simi-lar histological elements that the organs and tissues of the body are built up. But in one group, the *Infusoria*, the protoplasmic mass which constitutes the germ does not undergo this process of preliminary subdivision, but such structure as the adult animal possesses is the result of the direct metamorphosis of parts of its protoplasmic substance. Hence, morphologically, the bodies of these animals are the equivalents of a single cell; while, physiologically, they may attain a considerable amount of complexity.

The Infusoria abound in fresh and salt waters, and make their appearance in infusions of many animal and vegetable substances, their germs either being contained in the sub-stances infused, or being wafted through the air. Their diffusion is greatly facilitated by the property which many of them possess of being dried, and thus reduced to the condition of an excessively light dust, without the destruction of their vitality; while their rapid propagation is, in the main, due to their power of multiplying by division, with extraordinary rapidity, when duly supplied with nourishment. The majority are free and provided with numerous cilia by which they are incessantly and actively propelled through the medium in which they live; but some attach themselves to stones, plants, or even the bodies of other animals. A few are parasitic, and the bladder and intestines of the Frog are usually inhabited by several species of large size.

The Bell-animalcules are Infusoria which are fixed, usually by long stalks, to water-plants, or, not unfrequently, to the limbs of aquatic Crustacea. The body has the shape of a wine-glass with a very long and slender stem, provided with a flattened disc-like cover. What answers to the rim of the wine-glass is thickened, somewhat everted, and richly ciliated, and the edges of the disc are similarly thickened and ciliated. Between the thickened edge of the cover, or *peristome*, and the edge of the disc, is a groove, which, at one point, deepens and passes into a wide depression, the *vestibulum*. From this a narrow tube, the *œsophagus*, leads into the central substance of the body, and terminates abruptly therein; and when fæcal matters are discharged, they make their way out by an aperture which is temporarily formed in the floor of this vestibule. The outermost layer of the substance of the body is denser and more transparent than the rest, forming a *cuticula*. Immediately beneath the cuticle it is tolerably firm and slightly granular, and this part is distinguished as the *cortical* layer; it passes into the central substance, which is still softer and more fluid.

In the undisturbed condition of the Bell-animalcule, the stem is completely straightened out; the peristome is everted, and the edges of the disc separated from the peri-

stome; the vestibule gaping widely and the cilia working
vigorously. But the least shock causes the disc to be retracted, and the edge of the peristome to be curved in and
shut against it, so as to give the body a more globular form.
At the same time, the stem is thrown into a spiral, and the
body is thus drawn back towards the point of attachment.
If the disturbing influence be continued, this state of retrac-
tion persists; but if it be withdrawn, the spirally coiled stem
slowly straightens, the peristome expands, and the cilia
resume their activity.

In the interior of the body, immediately below the disc,
a space, occupied by a clear watery fluid, is seen to make
its appearance at regular intervals—slowly enlarging until
it attains its full size, and then suddenly and rapidly dis-
appearing by the approximation of its walls. This is the
contractile vesicle. Whether it has any communication with
the exterior or not and what is its function, are still open
questions. If the Bell-animalcule is well fed, one or more
watery vesicles of a spheroidal form, each containing a certain portion of the ingested food, will be seen in the soft
central mass of the body. And by mixing a small quantity
of finely divided carmine or indigo with the water in which
the *Vorticella* live, the manner in which these food-vesicles
are formed may be observed. The coloured particles are
driven into the vestibule by the action of the cilia of the
peristome and the adjacent parts, and gradually accumulate
at the inner end of the gullet. After a time the mass here
heaped together projects into the central substance of the
body, surrounded by an envelope of the accompanying
water; and then suddenly breaks off, as a spheroidal drop,
henceforward free in the soft central substance. In some
Bell-animalcules, the food-vesicles thus formed undergo a
movement of circulation, passing up one side of the body,

then crossing over below the disc and descending on the other side. Sooner or later the contents of these vesicles are digested, and the refuse is thrown into the vestibule by an aperture which exists only at the moment of extrusion of the fæces, and is indistinguishable at any other time.

A portion of the substance of the body, which is slightly different in transparency and in its reactions to colouring substances from the rest, is called the *nucleus* or *endoplast*. It is elongated and bent upon itself into a crescentic or horseshoe shape.

The Bell-animalcules multiply in two ways; partly by *longitudinal fission*, when a bell becomes cloven down the middle, each half acquiring the structure previously possessed by the whole; and partly by *gemmation* from the *endoplast*, in which latter case the endoplast divides and one or more of the rounded masses thus separated are set free as locomotive germs.

Sometimes a rounded body, encircled by a ring of cilia but having otherwise the characters of a *Vorticella* bell, is seen to be attached to the base of the bell of an ordinary *Vorticella*. It was formerly supposed that these were buds, but it appears that they are independent individuals, which have attached themselves to that to which they adhere and are gradually becoming fused with it, so that the two will form one indistinguishable whole. It is probable that this "conjugation" has relation to a sexual process.

Under certain circumstances a *Vorticella* may become *encysted*. The peristome closes and the bell becomes converted into a spheroidal body, in which only the nucleus and the contractile vesicle remain distinguishable. This surrounds itself with a structureless envelope or *cyst*, from which, after remaining at rest for a longer or shorter time, the Bell-animalcule may emerge and resume its former state of existence. In thus passing into a temporary condition of rest many of the other Infusoria resemble *Vorticella*.

The two genera of Infusoria which most commonly occur in the Frog are *Nyctotherus* and *Balantidium*. Both are free and actively locomotive, and the former is particularly remarkable for its relatively large size and semilunar contour, and for the length and distinctness of its curved œsophagus. *Balantidium* is pyriform, and has a very short œsophageal depression.

LABORATORY WORK.

A. Examine duckweed roots, confervæ, &c., with ¼ inch objective avoiding pressure; having found a group of *Vorticellæ* note the following points with a higher power.

   1. **In the extended state of the animal.**

      a. *The body.*

         a. Its size (measure).

         b. Form; broadly speaking, that of an inverted bell : note—

            α. The prominent everted rim (*peristome*).

            β. The flattened central *disc* projecting above the peristome.

            γ. The *cilia* fringing the disc.

            δ. The depression between the peristome and disc.

            ε. The mouth of the chamber (*vestibulum*) into which the œsophagus and anus open, in the hollow between the peristome and disc.

c.   Structure.

   α.   The thin, transparent, homogeneous external layer (*cuticle*).

   β.   The granular layer (*cortical layer*) inside the cuticle.

      [Its fine transverse striation.]

   γ.   The central more fluid part, not sharply marked off from β.

      The various clear spaces (*alimentary vacuoles*) in it, containing foreign (swallowed) bodies (Diatoms, Protococcus, &c.).

   δ.   The *contractile vesicle;* its position, in the cortical layer just beneath the disc; its systole and diastole.

   ε.   The *nucleus;* an elongated curved body in the cortical layer; sometimes nearly homogeneous, sometimes more distinctly granular. The nucleus is usually indistinguishable until after treatment with iodine (4).

   ζ.   The *gullet;* sometimes seen in optical transverse section as a clear round space; sometimes seen sidewise as a canal opening above on the disc, and ending abruptly below in the body-substance.

b.   *The stalk.*

   α.   Its length and diameter (measure).

   β.   Its structure; the external homogeneous layer (*sheath*) continuous with the cuticle; the highly refractive centre (*axis*) generally surrounded with granules, and continuous with the cortical layer of the bell.

2. **In the retracted state.**

   a. *The body.*

      a. Its form; pear-shaped; rounded off above; no disc or peristome visible.

      β. The clear transverse space near the top, indicating the interval between the retracted disc and the rolled-in peristome. In this space the cilia can frequently be seen moving.

      γ. Structure; as in 1. a. c.

   b. The stalk; thrown into corkscrew-like folds.

3. **The movements of Vorticella.** Compare especially the regularity, definiteness and rapidity of some of them with the slow and irregular movements of *Amœba.* (III.)

   a. *The ciliary movement.*

      a. Examine the cilia carefully; delicate homogeneous processes; their length, diameter and form; their position.

      [β. The continuity of the cilia with the cortical layer.]

      γ. The function of the cilia; their rapid movements, alternately bending and straightening: the *co-ordination* of these movements; they work in a definite order; note the currents produced in the neighbouring water (if necessary introduce a few particles of carmine under the coverslip); the sweeping of small bodies down the gullet.

   b. *The movements of the contractile vesicle* (see III. A. 3. c). Tolerably regular rhythmic distension and collapse (diastole and systole).

c. *The currents in the central parts* of the body carrying round the swallowed bodies. (Compare VI. C.)

d. *The movements of the animal as a whole.* ($\frac{1}{2}$ inch or $\frac{1}{4}$ inch obj.)

    α. Its extreme *irritability;* it contracts on the slightest stimulation : often without any apparent cause.

    β. The movements which occur in contraction; the coiling up of the stalk ; the rolling in of the disc. The rapidity of these movements.

    γ. The mode of re-expansion ; the stalk straightens first ; then the peristome is everted; finally the disc and its cilia are protruded.

4. Stain with iodine or magenta; the cuticle uncoloured —the rest stained ; the nucleus especially becomes deeply coloured.

5. Treat with acetic acid ; the contents soon disappear (except perhaps some swallowed bodies)—the cuticle later or not at all.

6. Note the following points in various specimens—

    α. *Multiplication by fission;* a bell partially divided into two by a vertical fissure starting from the disc.

    β. Two complete bells on one stalk ; the result of completion of the fission. The development of a basal circlet of cilia by one or both of these bells.

    [γ. Free swimming unstalked bells (detached bells from β).]

[δ. *Conjugation;* the attachment of a small free swimming bell to the side of a stalked one.]

[ε. *Encystation;* the body contracted into a ball and surrounded by a thickened structureless layer, the contractile vesicle being persistently dilated.]

B. Other forms closely allied to *Vorticella* which may be met with, and which will do nearly as well for examination, are ;—

   a. **Epistylis.** Bell-shaped animals growing on a *branched non-contractile* stalk.

   b. **Carchesium.** A form very like *Vorticella* but borne on a *branched* contractile stalk.

   c. **Cothurnia.** An almost sessile form, provided with a cup or envelope into which the bell can be retracted.

[The activity of the movements of the free Infusoria interferes with the complete examination of the living animal. It is well therefore to add a little osmic acid solution to the drop of water under examination. This kills such Infusoria as *Paramæcium, Nyctotherus* and *Balantidium* instantly, without destroying the essential features of their organization.]

X.

THE FRESHWATER POLYPES (*Hydra viridis and H. fusca*).

If a waterweed, such as duckweed, from a pond, is placed in a glass and allowed to remain undisturbed for a short time, minute gelatinous-looking bodies of a brownish or green colour may frequently be found attached to it, or to the sides of the glass. They have a length of from $\frac{1}{4}$ to $\frac{1}{2}$ of an inch, and are cylindrical or slightly conical in form. From the free end numerous delicate filaments, which are often much longer than the body, proceed and spread out with a more or less downward curve, in the water. If touched, these threads, which are the *tentacles*, rapidly shorten and together with the body shrink into a rounded mass. After a while, the contracted body and the tentacles elongate and resume their previous form. These are *Polypes*, the brown ones belonging to the species termed *Hydra fusca*, the green to that called *H. viridis*. The polypes usually remain attached to one spot for a long time, but they are capable of crawling about by a motion similar to that of the looping caterpillar; and, sometimes, they detach themselves and float passively in the water.

When any small animal, such as a water-flea, swimming through the water comes in contact with the tentacles, it is grasped, and conveyed by their contraction to the aperture of the wide mouth, which is situated in the middle of the circle formed by the bases of the tentacles. It is then taken into a cavity which occupies the whole interior of the body; the nutritive matters which it contains are dissolved out and absorbed by the substance of the *Hydra;* and the innutritious residuum is eventually cast out by the way it entered. Small pieces of meat, brought within reach of the tentacles, are seized, swallowed and digested in the same manner.

If a *Hydra* is well fed, bud-like projections make their appearance upon the outer surface of the body. These gradually elongate and become pear-shaped. At the free end a mouth is formed; and around it minute processes are developed and grow into tentacles; and thus a young *Hydra* is formed by gemmation from the parent. This young *Hydra* becomes detached sooner or later, and leads an independent existence; but, not unfrequently, new buds are developed from other parts of the parent before the first is detached, and the progeny may themselves begin to bud before they attain independence. In this manner, temporarily compound organisms may be formed. Experiments have shewn that these animals may be cut into halves or quarters and that each portion will repair its losses, and grow up into a perfect *Hydra;* and there is reason to believe that this process of fission sometimes occurs naturally.

The *Hydra* multiplies by budding through the greater part of the year; but in the summer projections of the surface appear at the bases of the tentacles or nearer the attached. end of the body. Within the former (*testes*) great numbers of minute particles, each moved by a vibratile cilium, are developed and are eventually set free. Functionally, these answer to the antherozooids of plants, and they are termed *spermatozoa.*

7—2

The enlargement formed near the attached end of the polype may be single, as in *Hydra viridis,* or as many as eight may be found in other species. It becomes much larger than the testis, and is the *ovary.* Within it is developed a single large egg, or *ovum.* This ovum, which is a huge nucleated cell, is impregnated by the spermatozoa and undergoes division into two parts. Each of these again divides into two; and so on, until the ovum is broken up into a number of small embryo-cells. The mass of embryo-cells thus formed becomes surrounded with a thick, usually tuberculated or spinous, case; and, detaching itself from the body, forms the 'egg,' from which a new Hydra is developed.

Microscopic examination shews that the body of the *Hydra* is a sac, the wall of which is composed of two membranes, an outer (*ectoderm*), and an inner (*endoderm*). The tentacles are tubular processes of the sac, and therefore are formed externally by the ectoderm and lined internally by the endoderm. Both the endoderm and the ectoderm are made up of nucleated cells; the inner ends of those of the ectoderm being prolonged into delicate fibres, which run parallel with the long axis of the body on the inner face of the ectoderm. The green colour of the *Hydra viridis* results from the presence of chlorophyll grains imbedded in the protoplasm of the cells.

In both the ectoderm and the endoderm the protoplasm of the cells contains very singular bodies,—the so-called *urticating capsules, thread-cells,* or *nematocysts*—which are oval bags, with thick and elastic walls, containing a spirally coiled-up filament which is unrolled suddenly on the slightest pressure, and then presents the appearance of a long filament attached to the capsule, and often provided with three recurved spines near its base. As similar capsules of a larger size are the agents by which many of the jelly fishes sting severely, just as nettles do when they are handled, there is every reason to believe that the thread-cells of the *Hydra* exert a like noxious influence upon the small animals which serve as their prey.

Thus, *Hydra* is essentially a *cellular organism* like one of the lower plants, but differs from them morphologically in the fact that its cells are not enclosed within cellulose walls; and physiologically, in the dependence of these cells for their nutrition upon ready formed protein matter. The function of the chlorophyll granules contained in the endoderm of the green Hydra, and of the brown or orange-coloured particles in the endoderm of the other species, is wholly unknown.

The *Hydra*, again, may be compared to an aggregate of *Amœbæ*, which are arranged in the form of a double-walled sac and have undergone a certain amount of metamorphosis.

It is possible that the longitudinal fibres connected with the cells of the ectoderm may be specially contractile, and represent muscles; but, however this may be, each cell has its own independent contractility. No trace of a special nervous system has yet been discovered, and the manner in which the actions of the different parts of the *Hydra* are combined to a common end, as in locomotion and the seizing of prey, is not understood.

The *Hydra* has none of the special apparatuses which are termed sense-organs, or glands. The cavity of the body alone represents a stomach and intestine; there are no organs of circulation, respiration or urinary secretion; the products of digestion being doubtless transmitted, by imbibition, from cell to cell, and those of the waste of the cells exuded directly into the surrounding water.

LABORATORY WORK.

1. Put into a beaker some water containing bodies to
    which Hydræ are attached, and place the beaker in a
    window not exposed to direct sunlight : in the course
    of some hours many Hydræ will be found attached
    to that side of the glass which is turned towards the
    light.   Note their size, form, colour, mode of attach-
    ment and movements.

2. Transfer a Hydra, by means of a pipette, on to a
    slide; cover in plenty of water with a large coverslip,
    and examine with 1 inch obj.   Note—

a   **Form.**

   α. *The base* (so called *foot*): a flattened disc : nar-
       rower or wider than the body according to the
       state of extension of the latter.

   β. *The body proper:* cylindrical, varying much in
       length and diameter with the state of extension
       of the animal; its conical free end, with an open-
       ing (*mouth*) in it.   It is often difficult to see the
       mouth in this way, especially in the green species.
       It is readily seen however if a Hydra be placed
       in a drop of water, without a coverslip, and be
       watched with an inch objective until it turns its
       anterior end up towards the observer.

   γ. *The tentacles:* ranged round the mouth; their
       number and shape; their varying length and
       diameter ; the knob-like eminences on them.

   δ. *The testes:* small conical colourless eminences be-
       low the point of attachment of the tentacles.

ε. *The ovary:* a larger rounded colourless prominence near the base : there may be more than one.

ζ. *The buds:* young Hydræ, of various sizes and stages of development, attached to the sides of the parent.

Either δ, ε, or ζ, or all of them, may be absent in some specimens.

b. **Structure.**

α. The animal evidently composed of two layers, an outer, *ectoderm*, and inner, *endoderm*, the latter alone containing chlorophyll in the green species: the ectoderm is marked out into areas, and may with care be seen to be composed of distinct cells, though this is a little difficult to make out in fresh specimens.

β. The *body-cavity:* difficult to make out in the green species, frequently visible in the brown ones as a darker central patch with which the mouth-opening is continuous; the extension of the body-cavity into the tentacles. Note corpuscles floating along inside them when they are extended.

c. **Movements.**

α. The general *contractility* of the animal; it is constantly either extending or shortening its body and tentacles, and so altering its form and place.

β. Its *irritability;* slight pressure or other stimulus immediately causes it to contract.

3. Examine with a high power: try to make out the different cells of the ectoderm—

    a. Large somewhat conical nucleated cells, with the broader end turned outwards.

    β. Smaller rounded cells packed between the deep ends of the larger ones.

    γ. The *nematocysts:* small oval capsules, with a filament coiled up inside them, which are dispersed through the ectoderm in the interior of its component cells.

4. Treat with magenta : note the staining of the cells, the emission of the thread-cells, and the protrusion of their threads : three chief forms of thread-cell—

    a. An oval capsule with a filament many times its own length attached to one end, and three short processes radiating from the base of the thread.

    β. Smaller thread-cells, without the radiating processes and with a short thread.

    γ. Cells like *β*, but with a much longer thread.

5. Imbed in paraffin a Hydra which has been hardened in chromic or osmic acid[1] and cut sections from it ; or lay a prepared Hydra on a glass slide and with a razor cut off transverse slices ; having obtained by either method a number of thin sections mount them in glycerine and make out—

    a. The large and small cells of the ectoderm and its thread-cells, their arrangement and relations. (3).

---

[1] When a Hydra is placed in the above hardening fluids it nearly always contracts so much as to make it difficult to cut sections. If it be first killed, by placing it in a small quantity of water and when it has expanded adding some boiling water, fairly extended specimens for hardening can usually be obtained.

β. The cells of the endoderm: large, nucleated, with a flattened base and a rounded free end: their arrangement in a single layer.

γ. The thin intermediate layer (*muscular stratum*) between ectoderm and endoderm.

δ. The body cavity.

6. Tease out in water a specimen which has been treated with weak chromic acid (o. 1⅑) or with osmic acid: make out the various cells already described: notice branched tails proceeding from the narrower ends of the larger ectoderm cells.

[7. Tease out a fresh Hydra in water and observe the various cells. Note the amœboid movements exhibited by some, and the single cilium attached to other (endoderm) cells.]

8. Gently flatten out a testis in water by pressure on the coverslip, and examine with a high power. According to its state of maturity the following contents will be found in it—

a. A collection of the smaller ectoderm cells.

β. The same but having lost their nucleus and become hyaline.

γ. Cells otherwise like β, but with a long filament proceeding from them.

δ. *Ripe spermatozoa:* bodies consisting of a very small oval head to which a very delicate filament is attached, and which, should they get free, swim about in the water by the movements of this filament. They may frequently be seen in motion within the unruptured testis.

9.   Press out an ovary: according to its stage of development there will be found in it—

    α. Simply ectoderm cells with an unusual prepon-
derance of the smaller form.

    β. Imbedded among cells like *a*, one which has
become larger and clearer than the rest, and
possesses a distinct central clear spot in it.

    γ. Considerable aggregation of granular proto-
plasm round this cell, so as to form a body
consisting of a granular protoplasmic mass, in
which is imbedded a clear round vesicle, which
again contains a distinct rounded dot.

    δ. *The ripe ovum.*   Consisting of a great irregu-
larly branched mass of protoplasm (*vitellus*),
in which is a clear space (*germinal vesicle*)
containing another body (*the germinal spot*).

    ε. The *segmented ovum:* composed of a large
number of small cells.   Its thick capsule,
rough on its external surface.

XI.

THE FRESH-WATER MUSSEL
(*Anodonta Cygnæa*).

UNDER the name of 'Fresh-water Mussel' two distinct kinds of animals, which are not unfrequently abundant in our ponds and rivers, are included; namely, the *Anodonta* and two or three kinds of *Unio*. The *Anodonta* is chosen for special study here, but what is said about it applies very well to all parts of *Unio* except the shell.

The animal is enclosed in a shell composed of two pieces or *valves*, which are lateral, or right and left, in relation to the median plane of the body. The more rounded and broader end is anterior, the more tapering, posterior. If placed in a vessel of water, at the bottom of which there is a tolerably thick layer of soft mud or sand, and left quite undisturbed, the *Anodonta* will partially bury itself with its anterior end directed obliquely downwards; and the valves will separate at their ventral edges for a short distance. At the edges of this 'gape' of the shell the thickened margins of a part of the contained body which is called the *mantle*, become visible, and between them a large, whitish, fleshy, tongue-shaped structure—the *foot*—not unfrequently protrudes, and is used to perform the sluggish movements of which the *Anodonta* is capable. If some finely dividing colouring matter, such as indigo, is dropped into the water,

so as to fall towards the gape, it will be seen to be sucked
in; while, after a short time, a current of the same substance
will flow out from an opening between the two edges of the
mantle on the dorsal side of the posterior end of the body;
and these 'inhalent' and 'exhalent' currents go on, so long
as the animal is alive and the valves are open. Any disturb-
ance, however, causes the foot, if it was previously protruded,
to be retracted, while the edges of the mantle are drawn in
and the two valves shut with great force. On the other
hand, in a dead *Anodonta* the valves always gape, and if
they are forcibly shut spring open again. The reason of this
is the presence of an elastic band, which unites the dorsal
margins of the two valves, for some distance, and is put on
the stretch when the valves are forcibly brought together.
During life they are thus *adducted* by the contraction of two
thick bundles of muscular fibres, which pass from the inner
face of one valve to that of the other, one at the anterior
and the other at the posterior end of the body, and are called
the *anterior* and *posterior adductors*.

The animal can be extracted from the shell without
damage, only by cutting through these muscles close to their
attachments.   It is bilaterally symmetrical, the foot proceeding from the middle of its ventral surface ; the mouth is
median and situated between a projection, which answers to
the under surface of the anterior adductor muscle, and the
superior attachment of the foot.   On each side of the mouth
are two triangular flaps with free pointed ends—the *labial
palpi*—and behind these, on each side, two broad, plate-like
organs, with vertically striated outer surfaces, are visible.
These are the *gills* or *branchiæ*.   In the dorsal region, the
integument is soft and smooth ; on each side, it is produced
into large folds, the lobes of the mantle or *pallium*, which
closely adhere to the inner surface of the valves of the shell,

and end, ventrally, in the thickened margins already mentioned. They pass into one another in front of the mouth;
at the sides, they are united with the dorsal edges of the
outer gill-plates; and, behind, they extend upwards and on
to the dorsal face of the body, before finally passing into one
another above, and in front of, the anus, which is small,
tubular, prominent, and median.   Thus the anus is inclosed
in a part of the cavity bounded by the two mantle lobes,
which is relatively small and shallow, and is termed the *cloacal
chamber;* while the gills, the foot, and the palps, hang down
into the relatively large *branchial* chamber which occupies
the space between the mantle-lobes for the rest of their
extent.   It is the prolongation of the margins of the former
cavity which gives rise to the tubular *anal siphon* seen in so
many Lamellibranchs; while the *ventral* or *branchial siphon*
is a similar prolongation of the margins of the branchial
chamber.   The dorsal siphon is the channel through which
the exhalent currents pass; the ventral, that for the inhalent
currents.

The currents are produced and kept up by the action of
the cilia which abound upon the gills.   The latter are perforated by innumerable small apertures, and the chambers
contained between the two lamellæ of which each gill is
formed, are in communication, above, with the cloacal
chamber.   The cilia work in such a way as to drive the
water in which the animal lives from the outer surface of
each gill towards its interior.   Hence the current which sets
from the branchial to the cloacal chamber.

The current of water which is thus continually drawn into
the branchial chamber carries with it minute organisms, *Infusoria*, Diatoms and the like, and many of these are swept
to the fore part of the branchial chamber, where they enter
the mouth, and are propelled by the cilia which line its cavity

into the alimentary canal. The latter presents a short and wide gullet, a stomach surrounded by hepatic follicles, a long intestine coiled upon itself, in a somewhat complicated manner, and, finally, a rectum, which lies in the middle line of the dorsal aspect of the body, traverses the pericardium and the heart which lies therein, and finally ends in the anus.

As the mouth is below and behind the anterior adductor and the rectum passes in front of and above the posterior adductor, it is clear that the alimentary canal, as a whole, lies between the two adductor muscles.

Digestion, that is solution of the proteinaceous and other nutritive matters contained in food, is effected in the stomach and intestine; and the nutritious fluid, thus formed, transudes through the walls of the alimentary cavity and passes into the blood contained in the blood-vessels which surround it. This blood is thence carried into a large sinus, which occupies the middle line of the body under the pericardium and between the *organs of Bojanus* (see Laboratory Work 5), and receives the greater part of the blood returning from all parts of the body. From this median *vena cava*, branches are given off to the gills and open into the extensive vascular network which those organs contain. From this, again, trunks lead towards the pericardium and open into one or other of the two auricles of the heart, which communicate by valvular apertures with the ventricle. The ventricle gives off two aortic trunks, one of which, the anterior, runs forwards in the middle line, above the rectum, while the other runs backwards, below the rectum. From these two aortæ branches are given off which divide into smaller ramifications for the different regions of the body, and for the viscera, and finally terminate in channels which answer to the capillaries of the higher animals.

The pericardial cavity, in which the heart is lodged, is situated in the posterior half of the dorsal region of the body. Through its thin dorsal wall, and, still better, when it is carefully laid open, the heart can be seen beating. The auricles contract, and, after them, the ventricle; the wave-like contraction of the latter being much the more easily visible. The lips of the auriculo-ventricular apertures are so disposed that the blood is impeded from flowing back into the auricles, when the ventricles contract, and is forced out, either forwards or backwards, through the two aortæ. From these it finds its way to the capillaries, and returns from them to the *vena cava;* whence it is carried, through the organs of Bojanus, to the branchiæ. Here it becomes purified of carbonic anhydride, and receives oxygen from the water in which the branchiæ are plunged; and it is finally brought back in an arterialized condition to the heart.

The heart is therefore systemic and propels aerated blood.

The majority of the vessels which convey the blood from the vena cava to the branchiæ, traverse the walls of the dark-coloured organs—the organs of Bojanus—which has already been mentioned; and it is probable that they here part with their nitrogenous waste matters—the organ of Bojanus, in all probability, playing the part of a kidney. The cavity of the organ of Bojanus communicates, on the one hand, with the pericardium, and, on the other, with the exterior, by an aperture which is situated close to the attachment of the inner gill to the walls of the body. Thus the cavity of the pericardium communicates directly with the exterior, though by a roundabout way. But it also communicates directly with the venous system, by sundry small apertures placed in the anterior part of its floor. Hence it must contain a mixture of blood and water.

The blood of the *Anodonta* is colourless, and contains colourless corpuscles, which resemble those of Man in structure and present the same Amœbiform movements.

The nervous system of the *Anadonta* consists of three pairs of yellow ganglia ; the *cephalic*, situated at the sides of the mouth ; the *pedal*, placed in the foot ; and the *parieto-splanchnic*, on the under face of the posterior adductor muscle. They are united by commissural cords which connect the cephalic ganglia with one another, and with the pedal and parieto-splanchnic ganglia, respectively. The only sense organs which have been discovered, are a pair of auditory vesicles, connected by nervous cords with the pedal ganglia.

The sexes are distinct. The *testes* and *ovaria* are similar in character, being racemose glands, which, in the breeding season, occupy a great part of the interior of the body. There is one gland on each side, opening by a minute aperture close to that of the organ of Bojanus.

The spermatozoa have minute, short, rod-like bodies, to which a long, filamentous, active *cilium* is attached, and, thrown off in enormous numbers, make their way out with the exhalent currents.

The ova are spherical, and the vitelline membrane is produced at one point into a short open spout-like tube, with a terminal aperture, the *micropyle*, through which, in all probability, the spermatozoon makes its entrance. When fully formed, multitudes of these ova pass out of the oviducal aperture and become lodged in the chambers of the gills, particularly the external gill, which is frequently completely distended by them. Here they are hatched, and give rise to embryos, which are so wholly unlike the parent *Anodonta*, that they were formerly thought to be parasites, and received the name of *Glochidium*. The embryo *Anodonta* is provided with a bivalve shell. Each valve has the form of an equilateral triangle united by its base with its fellow, by means of an elastic hinge, which tends to keep the two wide open. The apex of the triangle is sharply incurved, and is produced into a strong serrated tooth, so that when the valves approach, these teeth are directed towards one another. The mantle is very thin, and the inner surface of each of its lobes presents three papillæ, terminated by fine pencils of hair-like filaments. What appears to be the oral aperture is wide, and its margins are richly ciliated. There is a single adductor muscle and a rudimentary foot, from which one or two long structureless filaments, representing the *byssus* of the sea-mussel, proceed. These byssal filaments become entangled with one another and tend to keep the 'Glochidia' in their places.

After a time the larval *Anodontæ* leave the body of the parent, and attach themselves to floating bodies—very commonly to the tails of fishes—by digging the incurved points of their valves into the integument in the latter case, and holding on by them as if they were pincers. In this situation they undergo a metamorphosis; the gills are developed, the foot grows, the auditory vesicles become conspicuous in it, and the young *Anodonta* at length drops off and falls into its ordinary habitation in the mud.

LABORATORY WORK.

1. In the natural state of the animal only the shell or *exoskeleton* is visible, or this may be slightly open, and then the edge of the membrane lining it (the *mantle*) may be visible. Raise one valve of the shell, by separating the mantle from it with the handle of a scalpel, and then cutting through two strong bodies

(*the adductor muscles*), one at each end of the animal, which run from one valve of the shell to the other and prevent their separation. The two valves will now be united only by their *ligament.*

2. **General form and structure.**

a. In the animal now laid bare may be distinguished—

α. A *dorsal border* turned towards the hinge of the shell, and nearly straight.

β. A curved *ventral border*, opposite the dorsal.

γ. A wider *anterior end.*

δ. A narrower *posterior end.*

ε. A *right and left side.*

b. **The mantle or pallium.**

α. A bilobed semitransparent membrane, one lobe lining each valve of the shell.

β. The continuity of the two lobes on the dorsal side of the animal; their separation along most of its ventral side, where each forms a thick yellowish free border.

γ. The union of the two pallial lobes, for a short distance, towards the posterior part of their ventral border.

δ. The rudimentary *dorsal* and *ventral siphons*, separated from one another at the point of union γ and each marked out by a part of the mantle-edge covered by short hair-like processes: the dorsal siphon completely closed below and forming a narrow oval slit; the ventral siphon open below and continuous with the cleft between the ventral edges of the mantle-lobes.

ε. The *branchial or pallial chamber:* turn back the ventral edge of that mantle-lobe from which the shell has been removed: a chamber is thus exposed into which the ventral siphon, and the cleft continuous with it, lead.

ζ. The *cloacal chamber:* pass a probe through the dorsal siphon; it will enter a small chamber, separated from the pallial chamber by a partition which unites the hinder part of the two inner gills (c. β).

c. **The contents of the pallial chamber.**

α. *The foot:* a large, yellowish, somewhat plough-share-shaped mass, in the middle line; its apex directed forwards and ventrally, towards the front of the cleft between the mantle-lobes.

β. *The gills or branchiæ:* two lamellar bodies on each side of the foot, but reaching farther back than it does: the outer gill on each side, attached to the mantle-lobe; the inner, attached to the foot in front, but farther back, separated by a cleft from it; and behind the foot, united across the middle line with its fellow so as to form a partition separating the cloacal from the pallial chamber.

γ. *The labial palps:* a pair of small triangular processes on each side, in front of the gills and on the dorsal end of the anterior edge of the foot.

8—2

    δ.   *The mouth:* each labial palp is continuous with its fellow across the middle line, and between the lip-like ridges thus formed, lies the wide mouth-opening.

    d.   *The anterior and posterior adductor muscles:* if the reflected mantle-lobe be turned down again, the oval divided ends of the adductor muscles can be seen. They appear to perforate the mantle.

2.   Now remove the animal completely from its shell, by detaching the other mantle-lobe from the valve to which it is fixed, and cutting through the attachments of the adductor muscles to that valve. The thick dorsal border of the animal and the continuity of the mantle-lobes will now be more readily made out than they could be previously (2. **b.** β).

4.   **The heart.**

    a.   On the dorsal border of the animal is a clear space, where the mantle is very thin and covers-in a cavity filled with fluid. This cavity is the *pericardium,* and through its walls the heart can be seen beating.

    b.   Pin the Anodon out in water between two pieces of loaded cork, or paraffin, so that its dorsal border is upwards, a mantle-lobe spread over each bit of cork, and its foot and gills hanging down between the two pieces: then carefully cut away the dorsal side of the pericardium without injuring the heart.

    c.   *The heart* will now be exposed; it is a yellow-ish transparent sac, exhibiting regular contractions and composed of a median and two lateral chambers.

α. The *venticle*, or median chamber; an oval sac, from each end of which a large vessel (*anterior* and *posterior aorta*) is continued; running through the middle of the ventricle is seen part of the alimentary canal. All parts of the wall of the ventricle do not contract together; but a sort of wave of contraction passes, from one end of it to the other, like the peristaltic contraction of the intestine in one of the higher animals.

β. *The auricles;* one of these will be seen on each side if the ventricle be gently pushed out of the way: each is a somewhat pyramidal sac, continuous with the ventricle at the apex of the pyramid.

5. **The organs of Bojanus.**

a. Divide the alimentary canal at the posterior part of the pericardiac chamber and turn it and the heart forwards, so as to lay bare the floor of the pericardium. Running along the middle line of this floor will be seen a large blood-sinus, the *great vena cava;* on each side of this, the floor is formed by the roof of a transparent sac (*the non-glandular part of the organ of Bojanus*), through which is seen a dark brown mass (*the glandular part of the organ of Bojanus*).

b. At the extreme front end of the pericardiac floor, immediately under the point at which the intestine enters the cavity, will be found a pair of oval openings; pass into each a bristle, tipped with a small knob of sealing-wax to prevent it from perforating a passage for itself: the opening will be found to lead into a channel which runs along the glandular part of the organ of Bojanus.

c. Remove carefully the thin transparent roof of the non-glandular part of the organ of Bojanus, on one side, so as to lay bare the portion of the glandular part which lies within the non-glandular: the bristle will be found to leave the passage in the glandular portion by an aperture, which puts it in communication with the non-glandular part, and is situated on the upper side of the glandular part, opposite the posterior end of the pericardium. The glandular part extends back some way beyond this point; but it is imbedded closely in the neighbouring tissues, and is not contained in the loose non-glandular sac, which reaches back no farther than the posterior end. of the pericardium.

d. Examine the floor of the non-glandular part, at its anterior end: in it will be found a small aperture; gently push a guarded bristle through this: then turn the animal over, and detach the front end of the inner gill on the same side, from the foot. The bristle will be found to have passed out by an aperture (*external opening of the organ of Bojanus*) which lies just above the attachment of the gill to the body.

6. **The gills or branchiae.**

a. Cut out one of the gills and examine it; it will be found to consist of two lamellæ united by their ventral edges and enclosing a central cavity, which opens into a chamber (*epibranchial*) above, which is continued back to open into the cloacal chamber. The cavity between the lamellæ is subdivided by irregular partitions, which pass from one lamella to the other.

b. Carefully cut out a bit of the wall of the gill-sac on one side; mount in water and examine with 1 inch obj. The outer surface will be seen to be formed by parallel vertical bars, containing pairs of short rods; the inner face being formed by a meshwork of large vessels, perforated by wide apertures.

c. Examine with a higher power: the margins of each cleft will be found covered with large active cilia.

7. **The nervous system.**

a. The cerebral ganglia.

α. These will be found by carefully dissecting away the bases of the labial palps and the integument on the dorsal side of the mouth. They are two in number, each about the size of a pin's head, and somewhat triangular in form.

β. The *commissures* connected with the cerebral ganglia are—

A short cord uniting the two ganglia across the middle line over the mouth.

A cord, the *cerebro-pedal commissure*, which runs downwards and backwards from each and becomes continuous with that which runs forwards from the pedal ganglion of the same side (*b. β*).

A long slender cord which passes directly backwards from each beneath the organ of Bojanus and joins the parieto-splanchnic ganglia of the same side (*c*).

b.   *The pedal ganglia.*

   α.   Lay the animal on one side and proceed gently to scrape away the tissues of the foot at about the junction of its anterior with its middle third, where the muscular and the visceral portions of the foot join. The pedal ganglia will thus be brought into view. They are a pair of deep-orange-coloured oval bodies, each rather larger than a big pin's head; they are applied to one another in the middle line.

   β.   From each ganglion one *commissural cord* (*a. β*) passes forwards and upwards to the cerebral ganglion of its side, and branches are given off to the muscles of the foot and to the auditory organ.

c.   *The parieto-splanchnic ganglia.*

   α.   This pair are readily found by turning the animal on its dorsal side, and dissecting away the integument from the ventral surface of the posterior adductor muscle.

   β.   Trace forwards from each the cord (*a. β*) which runs to the cerebral ganglion of the same side. It is easy to follow the commissure so long as it lies in the region of the organ of Bojanus—difficult further on.

8. **The auditory organ.**

  a. This is rather difficult to dissect out in *Anodon:* it is a small sac which may be found by tracing back the posterior cord given off from the pedal ganglion, to a branch of which it is attached. There is usually an auditory vesicle connected with each pedal ganglion.

  b. If a fresh *Cyclas*[1] be obtained, and its foot removed, mounted in water, and examined with 1 inch obj., the auditory sac can readily be seen with a constantly-trembling particle, the *otolith*, in it.

9. **The alimentary canal.**

  a. This should be dissected out in another *Anodon* which has been well hardened in spirit. Carefully dissect away the thin layers of muscle which cover the left side of the foot: as this is done the dark-looking coil of the intestine will come into view: the two coils lying parallel to one another near the posterior border of the foot being probably those first seen. Continue to pick away the muscles and reproductive cæca until as much as possible of the course of the intestine is exposed. Make a small hole in it in one of the hindermost coils, pass in the end of a blow-pipe and inflate: then carefully lay open the intestine throughout its whole length so as to expose its inner surface; working towards the stomach on the one hand and the rectum on the other. Pass a guarded bristle into the mouth as far as it will readily go, and then lay open the

---

[1] *Cyclas cornea*—a small fresh-water lamellibranchiate mollusk.

alimentary canal along it, with a pair of scissors. Then push the bristle gently a little farther on, and follow it with the scissors, and so on, until the part where the intestine has already been laid open is reached.

b. The alimentary canal first runs towards the dorsal side for a short way (*œsophagus*), lying on the ventral side of the anterior adductor muscle : it then dilates into an irregular sac (*the stomach*); behind the stomach it continues as a long narrow tube, *the intestine;* this turns abruptly down, behind the stomach, into the foot, running at first towards its postero-inferior border; then curves up and forwards in the foot to near its dorsal part; then bends abruptly down and backwards again, parallel to its previous course, towards the ventral part of the foot, where it makes another turn and after running forwards some way turns upwards and runs to the anterior part of the pericardium, where it turns backwards and runs as a straight tube (*the rectum*), first through the ventricle of the heart, and then (passing on the dorsal side of the posterior adductor muscle) along the dorsal side of the cloacal chamber, in which it ends in an opening, *the anus*, placed on a prominent papilla.

c. On the sides of the stomach lies a brownish glandular mass, *the liver*.

   a. Tease out a bit of the liver in water, and examine with $\frac{1}{8}$ obj. It is composed of branched cæcal tubes lined by a layer of brownish epithelial cells.

10. **Reproductive organs.**

   a. The animals are diœcious, but the reproductive organs are similarly constructed in both sexes : they vary much in size with the season, being large in winter and spring, but small at other times.

   b. Close to the external opening of the organ of Bojanus will be found another small opening on each side, this is the *generative opening.*

   ç. From the generative opening can be traced back a duct, which divides into many cæcal branches which lie in the upper part of the foot.

11. **Muscular system.**

   a. This is most readily dissected out in a specimen which has been hardened in spirit. The chief muscles are :

      α. The *anterior* and *posterior adductor muscles* which pass directly from one valve of the shell to the other. These have already been seen.

      β. The *posterior retractor of the foot:* this can readily be found, on each side, running into the foot from its attachment to the shell in front of the posterior adductor muscle.

      γ. The *anterior retractor of the foot:* this runs from its attachment to the shell behind the anterior adductor muscle, into the front of the foot.

      δ. The *protractor of the foot* arises from the inner surface of the shell behind the organ

of the anterior adductor and below that of
the anterior retractor. Its fibres spread out in
a fan-like manner over the upper part of the
foot, some of them extending over the surface of the liver.

ε.   The *lesser retractors*.   Several very small
muscles arising from the shell just in front
of the umbo and spreading over the surface
of the liver.

ζ.   The *intrinsic foot-muscles*: forming the
greater part of the ventral portion of that
organ.

η.   Small *muscles* attached to each mantle-lobe,
at some little distance from its swollen free
edge and fixed to the shell along a linear
impression, which runs from one adductor to
the other and is termed the *pallial line*.

b.   Tease out in glycerine a bit of one of the muscles which has been treated with 0·5% chromic
acid solution.   Examine with $\frac{1}{8}$ inch obj.   It
is composed of spindle-shaped flattened cells,
in each of which lies an elongated nucleus:
the substance surrounding the nucleus is clear,
but the rest of the cell is granular and contains a great number of small particles arranged
pretty definitely in transverse rows. While these
muscular fibres agree in form with those of
smooth muscles, in minute structure they approach *striped muscles*.

12.   **The shell or exoskeleton.**

a.   Its two hardened lateral pieces or valves; each
with a straight dorsal and a curved ventral edge,

and .an anterior larger and posterior smaller end : note the soft uncalcified ventral edge of each valve.

b. The *umbo;* a small blunt eminence on the dorsal border of each valve near its anterior end.

c. The *ligament:* an elastic uncalcified part of the exoskeleton behind the umbones, uniting the two valves and tending to keep their ventral edges slightly separated.

d. *The markings on the shell.*

   a. *External markings.* The outside of the shell is greenish brown, and on it are seen a number of concentric lines generally parallel to the margin of the shell, and more numerous towards the ventral edge.

   β. *Internal markings.* The interior of the valve is white and iridescent : on it are seen, near the dorsal border, two oval marks, the *anterior* and *posterior adductor impressions.*

     Joining the two adductor impressions is a curved line, the *pallial impression,* which marks where the muscles of the edge of the mantle were fixed to the shell.

     In front of the posterior adductor impression is seen a small mark, indicating where the posterior retractor muscle was fixed.

     Behind the anterior adductor impression are two marks, one opposite its upper, the other opposite its lower end : the former indicates the point of attachment of the *anterior retractor*, the latter of the *protractor pedis* muscle.

Extending from each adductor impression towards the umbo is a fainter, gradually tapering impression, which may be followed into the cavity of the umbo, and indicates the successive attachments of the adductor muscles, as the animal has increased in size.

13. In the breeding season, examine the contents of the testis for spermatozoa, and those of the ovary for ova. Note the micropyle of the latter. If the outer gill appear to be thick and distended, it will be found full of the larvæ of the *Anodon,—Glochidium.* Note the characters of their shells and the entangled filaments, or *byssus,* with which they are provided.

XII.

THE FRESH-WATER CRAYFISH (*Astacus fluviatilis*) AND THE LOBSTER (*Homarus vulgaris*).

THE Crayfish and the Lobster are inhabitants of the water, the former occurring in many of our rivers and the latter abounding on the rocky parts of the coasts of the European seas. They are bilaterally symmetrical animals, provided with many pairs of limbs, among which the large prehensile 'claws' are conspicuous. They are very active, walking and swimming with equal ease and sometimes propelling themselves backwards or forwards, with great swiftness, by strokes of the broad fin which terminates the body. They have conspicuous eyes, mounted upon moveable stalks, at the anterior end of the head; and two pairs of feelers, one pair of which are as long as the body, while the other pair are much shorter.

The body and limbs are invested by a strong jointed shell, or *exoskeleton*, which is a product of the subjacent epidermis, and consists of layers of membrane which remain soft and flexible in the interspaces between the segments of the body and limbs, but are rendered hard and dense elsewhere by the deposit of calcareous salts; the exoskeleton is deeply tinged with a colouring matter which turns red when exposed to the action of boiling water. The body presents an anterior division—the *cephalothorax*—covered

by a large continuous shield, or *carapace;* and a posterior
division—the *abdomen*—divided into a series of segments
which are moveable upon one another in the direction of
the vertical median plane, so that the abdomen can be
straightened out, in *extension;* or bent into a sharp curve, *in
flexion.* Of these segments there are seven. The anterior
six are the *somites* of the abdomen, and each of them has a
pair of appendages attached to its ventral wall. The seventh
bears no appendages and is termed the *telson.* The anus is
situated on the ventral aspect, beneath the telson and behind
the last somite.

A groove on the surface of the carapace, which is termed
the *cervical suture,* separates an anterior division, which
belongs to the head or *cephalon,* from a posterior division
which covers the *thorax;* and the thoracic division of the
carapace further presents a central region, which covers
the head, and wide lateral prolongations, which pass downwards and cover the sides of the thorax, their free ventral edges being applied against the bases of the thoracic
limbs. These are the *branchiostegites.* Each roofs over a
wide chamber in which the gills are contained and which
communicates with the exterior, below and behind, by the
narrow interval between the edge of the branchiostegite and
the limbs. Anteriorly and inferiorly, the branchial chamber
is prolonged into a canal, which opens in front and below
at the junction of the head with the thorax. In this canal
there lies a flat oval plate—the *scaphognathite*—which is
attached to the second pair of maxillæ and which plays a
very important part in the performance of the function of
respiration. Of the thoracic limbs themselves there are
eight pairs, and, on the ventral face of the body, the lines
of demarcation between the eight somites to which these
limbs belong may be observed. There is no trace of any

corresponding divisions in the cephalothorax of the Lobster; but, in the Crayfish, the last thoracic somite is incompletely united with those which precede it. The four posterior pairs of thoracic limbs are those by which the animal walks and are termed the *ambulatory legs.* The next pair is formed by the great claws or *chelæ.* The anterior three pairs are bent up alongside the mouth and are moved to and from the median line so as to play the part of jaws, whence they are termed foot-jaws or *maxillipedes.* The external or third pair of these maxillipedes are much stouter and more like the ambulatory limbs than the rest, and the inner edges of their principal joints are toothed. The innermost or first pair of maxillipedes are broad, foliaceous and soft. When these foot-jaws are taken away, two pairs of soft foliaceous appendages come into view. They are attached to the hinder part of the cephalon and are the jaws or *maxillæ.* The second, or outermost, is produced, externally, into the scaphognathite, which will be seen to lie in a groove which separates the head from the thorax laterally and is the *cervical groove.*

Anterior to these maxillæ lie the two very stout mandibles. Between their inner toothed ends is the wide aperture of the mouth, bounded, in front, by a soft shield-shaped plate, the *labrum;* and behind, by another soft plate, divided by a deep median fissure into two lobes, which is the *metastoma.* Thus far, the surfaces of the somites to which the appendages are attached look downwards, when the body is straightened out and the carapace is directed upwards. But, in front of the mouth, the wall of the body to which the appendages are attached is bent up, at right angles to its former direction, and consequently looks forwards. This bend of the ventral wall of the body is the *cephalic flexure.* In correspondence with this change of position of the sur-

M.   9

face to which they are attached, the three pairs of appendages of the somites which lie in front of the mouth are
directed either forwards, or forwards and upwards. The
posterior pair consists of the long feelers or *antennæ :* the
next, of the short feelers or *antennules;* and the most anterior
is formed by the short subcylindrical stalks (*ophthalmites*),
on the ends of which the eyes are situated.

This enumeration shews that the Lobster and Crayfish
have six pairs of abdominal appendages—the swimmerets;
eight pairs of thoracic appendages (four pairs of ambulatory
limbs, one pair of chelate prehensile limbs, three pairs of
maxillipeds), and six pairs of cephalic appendages (two pairs
of maxillæ, one pair of mandibles, one pair of antennæ, one
pair of antennules, one pair of eyestalks), making in all twenty
pairs of appendages. In correspondence with the number of
appendages the body consists of twenty somites; of which six
remain moveable upon one another to form the abdomen,
while the other fourteen are united to form the cephalothorax.

The branchiostegite is an outgrowth of the dorsolateral
region of the confluent thoracic somites. The serrated
rostrum which ends the carapace is a fixed median prolongation of the dorsal wall of the anterior cephalic somites;
while the telson is a moveable median prolongation of the
dorsal wall of the sixth abdominal somite. The labrum and
the metastoma are median growths of the sterna of the
præoral and post-oral somites.

Thus the whole skeleton in these animals may be considered as a twentyfold repetition of the ring-like somite
with its pair of appendages, which is seen in its simplest
form in one of the abdominal somites. Moreover, not-
withstanding the great variety of functions allotted to the
various appendages, the study of the details of their structure (see Laboratory work) will shew that they are all re-

ducible to modifications of a fundamental form, consisting of a basal joint (*protopodite*) with three terminal divisions (*endopodite, exopodite, epipodite*).

As has been already said, the Lobster and Crayfish are bilaterally symmetrical; that is to say, a median vertical plane passing through the mouth and anus divides them into two similar halves. This symmetry is exhibited not merely by the exterior of the body and the correspondence of the paired limbs, but extends to the internal organs; the alimentary canal and its appendages, the heart, the nervous system, the muscles and the reproductive organs, being disposed so as to be symmetrical in relation to the median vertical plane of the body.

The wide gullet leads almost vertically into the spacious stomach, and both are lined by a chitinous continuation of the exoskeleton. The stomach is divided by a transverse constriction into a spacious *cardiac*, and a much smaller *pyloric* division, from which latter the intestine passes. The walls of the anterior half of the cardiac sac are thin and membranous, but, in the posterior half, they become calcified so as to give rise to a *gastric skeleton* of considerable complexity. The chief part of this skeleton consists of a median dorsal T-shaped 'cardiac' ossicle, the cross-piece of which forms a transverse arch, while its long median process extends backwards in the middle line. The ends of the transverse arch are articulated obliquely with two small 'antero-lateral' pieces, the extremities of which again are articulated with postero-lateral pieces, and these unite with a cross-piece, the 'pyloric' ossicle, which arches over the roof of the pyloric division of the stomach. In this manner a sort of hexagonal frame with moveable joints is formed, and the median process projects backwards so far, as to end below the pyloric piece. It is connected with this, however, by a short 'pre-pyloric' ossicle which ascends obliquely forwards and is articulated with the anterior edge of the pyloric piece. The lower extremity of this is produced into the strong median 'uro-cardiac' tooth ; two small 'cardiac' teeth are borne by the median process of the cardiac ossicle; while the postero-lateral pieces are flanged inwards, and, becoming greatly thickened and ridged, form the large 'lateral cardiac' teeth. Two powerful muscles are attached to the cardiac ossicle, and ascend obliquely forwards to be inserted into the under face of the carapace. Two other similar muscular bundles arise from the pyloric ossicle, and, passing obliquely upwards and backwards, are also inserted into the under face of the carapace. The disposition of all these parts is such that when these muscles contract, the uro-cardiac tooth moves forwards and downwards, while the lateral teeth move inwards downwards and backwards, and the three meet in the middle line. The action of these muscles can be readily imitated by seizing the anterior and posterior cross-pieces with forceps and pulling them in the direction in which the muscles act. The three teeth will then be seen to come together with a clash. Thus the food which has been torn by the jaws is submitted to further crushing in this gastric mill. The walls of the pyloric division of the stomach are thick, and project like cushions into its interior, thereby reducing its cavity to a narrow passage. The cushion-like surfaces of the pyloric walls are provided with long hairs which stretch across this narrow passage, and thus convert it into a strainer, which allows of the passage of only very finely divided matter from the gastric sac to the thin and delicate intestine. The hepatic ducts open, one on each side, at the junction of the pyloric division of the stomach with the intestine. The

intestine is slender and delicate, smooth internally in the
Lobster, papillose in the Crayfish.   Near its hinder end its
walls become thicker for a short distance, and this thickened portion, with which, in the Lobster, a short dorsal
cæcum is connected, may be regarded as the large intestine
or rectum.

The heart is a short, thick, somewhat hexagonal, symmetrical organ lodged in the pericardiac sinus, to the walls of which
it is attached by fibrous bands.   In its anterior half three
pairs of apertures are visible, two being placed upon the
upper face, two at the sides, and two on the under face.   The
lateral apertures are the most posterior, the dorsal, the most
anterior in position.   Each aperture begins in a funnel-
shaped depression of the outer face of the organ, which leads
obliquely inwards and terminates by a valvular slit in the
cavity of the heart.   This cavity is very much reduced by
the encroachment of the muscular bands which constitute
the walls of the heart, so that a transverse or longitudinal
section shews only a small median cavity surrounded by a
thick and spongy wall.

During life, the heart beats vigorously, the whole of its
parietes contracting together.   From the dorsal part of its
anterior extremity three arteries are given off, one median
and two lateral, to the cephalon and its contents, and from
the ventral aspect of this end of the heart an *hepatic artery*
is given off, on each side, to the liver.   At its posterior end,
the heart ends in a median dilatation from which two great
arterial trunks are given off; one the *superior abdominal*
artery, which runs along the dorsal face of the intestine,
giving off transverse branches as it goes, in each somite; and
the other, the *sternal* artery, which passes ventrally to the
interspace between the penultimate and antepenultimate

thoracic ganglia, passes between their commissures and divides into two branches, which run, backwards and forwards, between the ganglionic chain and the exoskeleton.

These arteries divide and subdivide and end in what, in some parts of the body at any rate, *e.g.* the liver, is a true capillary system. The veins are irregular channels, or *sinuses*, which lie between the several muscles and viscera. One of the largest of these is situated in the median ventral line, and can be readily laid open by piercing the soft integument which lies between any two of the abdominal sterna. The blood flows out of the aperture with great rapidity, and the quantity shed shews the size of the sinus and its free communication with the rest of the vascular system. By cutting across any one of the limbs and inserting a blow-pipe into the place whence the blood wells forth, this ventral sinus can be readily injected with air. A large and irregular sinus is also to be found in the median dorsal region of the abdomen and is freely connected with the median ventral sinus. The stem of each branchia contains two canals, one running along its outer and the other along its inner face. The outer canal communicates, at its origin, with the median ventral sinus. The inner canal opens into a passage which ascends in the lateral wall of the thorax and opens, after meeting with other '*branchio-cardiac*' canals, opposite the lateral aperture of the heart. As the valvular lips of this and the other apertures of the heart open inwards, the blood, when the systole takes place, is driven out of the heart through the various arteries, and a considerable part of the blood thus propelled into the capillaries is collected by the median ventral sinus and thence, passing through the gills, eventually returns to the heart, which is therefore, like the heart of *Anodon*, a *systemic* and not a *branchial* heart. But whether the whole of the venous blood takes the same

course, or whether some of it returns from the dorsal sinuses
directly to the pericardium, is a question which is not de-
cided.   Nor is it certain whether the so-called pericardium
is to be regarded as one cavity, or whether the fibrous bands,
which connect the heart with its walls, may not subdivide it
into compartments in immediate communication with certain of the cardiac apertures, and not with the rest.

In the Lobster, from which the blood is readily obtained
in quantity, it is a nearly colourless fluid, which usually has
a faint neutral tint.   It readily coagulates, a tolerably firm
clot separating from the serum.   It contains nucleated corpuscles, devoid of any noticeable colour, which throw out
very long pseudopodial prolongations, and thereby take an
irregularly stellate form.

It has been seen that the respiratory organs, or branchiæ,
are lodged in a chamber situated between the branchiostegite
externally, the lateral walls of the thoracic somites internally,
and the bases of the thoracic limbs below; and that there
is a narrow interspace between the free edge of the branchiostegite and the latter.   At the anterior end of the chamber, a funnel-shaped passage leads to the anterior opening
mentioned above, and, in this passage, the scaphognathite
lies like a swing door.

During life, the scaphognathite is in incessant movement
forwards and backwards, scooping out the water in the branchial chamber through its anterior aperture at every forward
motion.   The place of the water thus thrown out is taken by
water which flows in by the inferior and posterior cleft beneath the free edge of the branchiostegite, and thus a constant
current over the gills is secured.   Each branchia is somewhat
like a bottle-brush, having a stem beset with numerous filaments; and the blood contained in the vessels of the latter

being separated by only a very thin membrane from the air contained in the water, loses carbonic anhydride and gains a corresponding amount of oxygen in its course through the branchiæ.

The branchiæ are attached partly to the epimera of the thoracic somites, partly to the proximal ends of the thoracic limbs. The epipodites of the limbs ascend between the sets of branchiæ which belong to each somite, and separate them. The branchiæ which are attached to the limbs must necessarily be stirred by the movement of the latter, and hence the exchange of gases between the blood which they contain, and the water must be, to a certain extent, increased, in proportion to the muscular contractions which give rise to the movements of the limbs and the consequently increased formation of carbonic anhydride.

The mode and place of the excretion of nitrogenous waste is not yet clearly made out, but it seems probable that two large green glands which lie in the cephalon, close to the bases of the antennæ, are renal organs. Each gland encircles the neck of a large thin-walled sac which opens by a short canal upon the ventral face of the basal joint of the antenna.

The nervous system consists of a chain of thirteen ganglia—united by longitudinal commissures—lodged in the median line of the ventral aspect of the body, from which nerves are given to the organs of sense, to the muscles of the trunk and limbs, and to the integuments; and of a *visceral* nervous system, developed chiefly upon the stomach.

Of the thirteen ganglia, the most anterior lies in the cephalon, close to the attachments of the three anterior pair of appendages, and gives branches to them and to the visceral nervous system. It is usually termed the brain or the *supraœsophageal* ganglion. It is connected by two commissural cords, which pass on each side of the gullet, with a larger ganglionic mass, which is called the *subœsophageal* ganglion.   This occupies the region of the hinder part of the cephalon and the anterior part of the thorax, and gives off nerves to the maxillæ and the three pair of maxillipeds. Five other ganglia lie in the five somites which bear the chelæ and the ambulatory limbs, and there is one for each abdominal somite, the last of these being the largest of the six.

The longitudinal commissures between the abdominal ganglia are single ; but, in the thorax, the commissures are double, and the ganglia themselves shew more or less evident indications of being double.   And there is reason to believe that these thirteen apparent ganglia really represent twenty pairs of primitive ganglia, one pair for each somite ; the three pairs of præoral ganglia having coalesced into the brain ; and the five which follow the mouth having united into the subœsophageal mass.

The only organs of special sense which are recognizable in the Lobster and Crayfish are eyes and auditory organs.

The *eyes* are situated at the extremities of the eyestalks, or ophthalmites, which represent the first pair of appendages of the head.   The rounded end of the eyestalk presents a clear, smooth area of somewhat crescentic form, divided into a great number of small four-sided facets.   This area corresponds with the *cornea*, which is simply the ordinary chitinous layer of the integument become transparent. The inner face of each facet of the cornea corresponds with the outer end of an elongated transparent slightly conical body—the *crystalline cone*—the inner end of which passes into a relatively long and slender *connective rod*, by which it is united with a spindle-shaped transversely striated body—

the *striated spindle.* The inner extremity of this again is connected with the convex surface of the dilated cushion-shaped ganglionic termination of the optic nerve. The respective *striated spindles, connective rods* and *crystalline cones,* thus radiate from the outer surface of the terminal ganglion to the inner surface of the cornea, and each is separated from its neighbour by a nucleated *sheath,* parts of which are deeply pigmented. Nothing is accurately known as to the manner in which the function of vision is performed by the so-called *compound eye* which has just been described. The inner and outer faces of the corneal facets are flat and parallel. They therefore cannot play the part of lenses; and, if they could, there is no trace of nerve endings so disposed as to be affected by the points of light gathered together in the foci of such lenses. Morphologically, the cones, connective rods and striated spindles, are in many ways analogous to those elements of the retina of the Vertebrata which make up the layers of rods and cones and the granular layers. These structures are properly modifications of the epidermis; inasmuch as the cerebral vesicle, of which the retinal vesicles are outgrowths, are involutions of the epidermis of the embryo, and, morphologically speaking, the free ends of the rods and cones of the vertebrate eye are, as in the crustacean, turned outwards. It seems probable, therefore, that the crustacean eye is to be compared to the retina alone of the vertebrate eye, and that vision is performed as it would be by the retina deprived of its refractive adjuncts.

The *auditory organ* of the Lobster and Crayfish is situated in the basal joint of the antennule, on the dorsal surface of which a small slit-like opening, protected by numerous hairs, is to be seen. The chitinous layer of the integument is invaginated at this opening, and thus gives rise to a small flattened sac lodged in the interior of the antennule. One side of this sac is in-folded so as to produce a ridge, which projects into the cavity of the sac, and is beset with very fine and delicate hairs. The auditory nerve enters the fold, and its ultimate filaments reach the bases of these hairs. The sac contains water in which minute particles of sand are suspended.

The sexes are distinct in the Lobster and Crayfish. The external characters of the males and females and the form of the reproductive organs are described in the Laboratory work.

The impregnated ova are attached in great numbers, by a viscid secretion of the oviduct, to the hairs of the swim-merets, where they undergo their development. A Lobster with eggs thus attached, is said by the fishermen to be 'in berry.' In the Crayfish, the embryo passes through all the stages which are needed to bring it very near to the form of the adult before it leaves the egg: but, in the Lobster, the young, when hatched, are larvæ extremely un-like the parent, which undergo a series of metamorphoses in order to attain their adult condition. The larvæ may fre-quently be obtained by opening the eggs of a 'hen-lobster' in 'berry.' They have a rounded carapace, two large eyes, a jointed abdomen devoid of appendages; and the thoracic limbs are provided with long exopodites.

The ordinary growth, no less than the metamorphoses of the Lobster and Crayfish, are accompanied by periodical castings of the outer, chitinous, layer of the integument. After each such *ecdysis*, the body is soft and the animal retires into shelter until the 'shell' is reproduced.

LABORATORY WORK.

1. **General external characters.**

The animal is covered by a dense *exoskeleton:* in it are readily recognised the following parts :—

a. The *body proper :*

   α. Its anterior unsegmented portion (*cephalotho-rax*): the great shield-like plate (*carapace*) covering the back and sides of the cephalotho-rax; the groove across the carapace (*cervical suture*) marking out the line of junction of *head* proper and *thorax :* the anterior pro-longation of the carapace to form the *frontal spine.*

   β. The posterior segmented portion (*abdomen*): its seven divisions; the anterior six much like one another; the most posterior (*telson*) different from the rest.

b. The great number of jointed limbs (*appendages*) attached to the ventral aspect of the body: their varying characters in different regions.

c. The external apertures of the body.

   α. The *mouth;* seen by separating the append-ages beneath the head.

   β. The *anus;* a longitudinal slit beneath the telson.

   γ. The *paired genital openings:* in the male, on the first joints of the last pair of appendages of the thorax: in the female, on the first joints of the last thoracic appendages but two.

[δ.   The *openings of the auditory organs.*

ε.   The *openings of the green glands.*

These will be more readily found when the appendages on which they are situated have been separated.   See 21. *f* and *g.*]

2.   Examine carefully the third abdominal segment or *somite* and its appendages.

    a.   *The segment proper:* arched above ; flattened below.

        α.   Its dorsal part (*tergum*), with an anterior smooth portion overlapped by the preceding segment in extension of the abdomen, and a posterior rougher part overlapping part of the succeeding segment.

        β.   The ventral surface of the segment : united with the corresponding portions of the preceding and succeeding segments by a flexible membrane.

        γ   The point of union of the appendages with the somite.

        δ.   The *sternum:* that portion of the ventral surface of the somite which lies between the points of attachment of the appendages.

        ε.   The *epimeron:* the portion of the ventral surface which lies on each side external to the attachment of the appendage.

This region is very short and passes almost directly into the inner walls of the pleuron.

        ζ.   The downward extension (*pleuron*) of the lateral walls of the somite formed by the prolongation of the tergum and epimeron: the smooth facet on the anterior half of the pleuron where it is overlapped by the one in front.

b.   *The appendages* or *swimmerets:* one on each side: the structure of each—

    a.   The short two-jointed basal portion (*protopo-dite*), consisting of a shorter proximal and a longer distal piece.

    β.   The antero-posteriorly flattened elongated lamellæ attached to the distal joint of the protopodite, an inner (*endopodite*) and outer (*exopodite*).

3.   The fourth and fifth abdominal segments: closely resembling the third.

4.   The sixth abdominal segment: its modified appendages.

    a.   The protopodite: represented by a single short strong joint.   (In the lobster there is an incomplete basal joint.)

    β.   The exopodite and endopodite : wide plates fringed with setæ : the exopodite divided into two portions by a transverse joint.   ·

5.   *The telson.*

A flattened plate bearing no appendages : subdivided by a transverse joint (it is undivided in the lobster) : the membranous character of the greater part of the ventral surface of its anterior division.

The *tail-fin;* formed by the telson and the appendages of the sixth abdominal segment.

6.   The second abdominal segment.

Closely resembling the third in the female : in the male its appendages are modified : the protopodite and basal joint of endopodite much elongated, and the latter produced into a plate rolled upon itself so as to form a demicanal, concave inwards. (In the lobster the endopodite is produced inwardly, into an oval process.)

7.   The first abdominal segment : its appendages ; rudimentary in the female (it has only one instead of two terminal divisions in the lobster) : in the male consisting of a single plate rolled in upon itself. (In the lobster the single terminal division has the form of a flat scoop or a narrow spoon with its concave side turned inwards.)

8.   *The structure of the cephalothorax.*

a.   Note again the carapace, with its frontal spine and cervical suture.

β.   Turn the animal over and note the very narrow sterna between the points of attachment of the thoracic appendages.

The last thoracic somite is not completely ankylosed with the one in front, on the vertical side in the crayfish.   In the lobster it is.

γ.   Raise with a pair of forceps the free edge of the lateral part of the carapace which lies just over the bases of the thoracic appendages, and is termed the *branchiostegite:* note that it is formed by the large united pleura of the thoracic segments, and overlaps a chamber in which the gills lie.

9. Note the plane in which the sterna of the anterior three somites of the animal (marked out by their appendages) lie—it is nearly at right angles to the plane of the remaining sterna of the cephalothorax—so that their appendages are directed forwards instead of downwards.

10. Cut a vertical section of a piece of the exoskeleton which has been decalcified by lying in $1\frac{9}{8}$ chromic acid solution for a few days.

   a. It will be seen to be composed of a large number of parallel laminæ which are thicker towards the outer part. The laminæ are marked by ill-defined parallel lines which run perpendicular to the surface, and which give their edges a striated appearance. The outermost layer is more transparent than the rest and wants this striation.

   b. The epidermis lying beneath the innermost of the above laminæ is composed of ill-defined branched nucleated granular cells: the outermost giving off a large number of short processes which end in clubbed ends and penetrate a short way into the exoskeleton.

11. **The respiratory organs.** Remove now the branchiostegite on one side and examine the *gills:* they are 18 in number, arranged in two sets.

   α. Six are attached to the epipodites of some of the appendages (2nd and 3rd maxillipedes, chelæ, 1st, 2nd, and 3rd pair of ambulatory limbs).

   β. The remaining 12 are fixed to the sides of the body, and each consists of a central stem giving off a number of delicate filaments.

γ. Cut away the gills, noting the two large channels in the stem of each, and observe the cervical groove at the front of the gill-chamber with the scaphognathite (21. *d. a.*) lying in it.

[δ. In the lobster there are 20 gills on each side, arranged as in the crayfish, except that there are 14 on the side of the body.]

12. **Circulatory organs.** Immerse the animal in water with its ventral surface downwards: cut away carefully with a pair of scissors the dorsal part of the carapace which lies behind the cervical suture and that part of the wall of the thorax from which the gills have been removed.

A chamber (*the pericardial sinus*) is thus laid bare in which lies a polygonal sac, *the heart.*

a. The six openings from the sinus into the heart; two superior, two inferior, and two lateral: pass bristles into them. The *arteries* arising from the heart; five anterior, one (*ophthalmic*) single in the middle line, the others (*antennary* and *hepatic*) in pairs; one, *the sternal,* the largest of all, given off from the posterior end.

b. Cut away the terga of the abdominal somites and follow back the *superior abdominal branch* of the sternal artery, removing carefully the muscles which lie over it in the abdominal region. It will be seen as a transparent tube lying in the middle line on the intestine (14. *b.*), or in the female lobster separated from it anteriorly by the posterior ends of the two ovaries. It gives off branches from its upper side to the muscles over it, and also a pair of branches which run out laterally in the intervals between each pair of somites. In the sixth abdominal somite it terminates by splitting up into three or four large branches which pass in a radiating manner into the telson. On account of the small size of the crayfish this artery is difficult to dissect in it.

c.   The *sternal artery* presents an enlargement at its commencement just where the above branch arises from it. It then passes vertically downwards towards the ventral surface, passing on one side of the intestine. Its subsequent course must be followed later (15).

13. Reproductive organs.

These differ considerably in the crayfish and the lobster. They lie partly beneath the heart, which must therefore be removed or pushed on one side in order to see them. Both animals are unisexual.

a.   *Of the Crayfish.*

   α.   *The testis.* A trilobed yellowish mass: two of its lobes are larger than the third and pass forwards side by side in the middle line: the third lobe is directed backwards.

   β.   The two *vasa deferentia* arise just where the posterior lobe of the testis meets the two anterior. Each is narrow near the gland, but widens as it proceeds back from it, and becoming extremely convoluted, finally ends at the genital opening on its own side (1. *c.* γ.). Trace the course of the vas deferens on that side from which the thoracic wall has been removed (12).

γ. Tease out a bit of the testis in water, and examine with ⅛ obj. : it will be seen to be composed of sacculated tubes. In it or in the
vas deferens some of the *spermatozoa* may be
found : they are motionless and have the form
of nucleated cells provided with radiating processes.

δ. *The ovary* is a gland in shape and colour
very similar to the testis of the male. From
it two short oviducts arise and pass almost
directly downwards to the genital openings
(1. *c. γ.*).

b. *Of the Lobster.*

a. *The testes* are two long tubes which lie partly
in the thorax and partly in the abdomen.
Their posterior portions meet in the middle
line, but in front they diverge, and about one
fourth the length of each from its anterior end
a short transverse branch unites the two.

β. The *vas deferens* arises a little in front of the
middle of each testis and passes without con-
volutions towards the genital opening. Its
distal half is dilated.

γ. Tease out a bit of the testis in water and examine with ⅛ obj. for *spermatozoa*. They are
motionless, and consist of an elongated cell
from one end of which three rigid pointed
processes radiate.

δ. The *ovaries* of the lobster are also elongated
and lie partly in the thorax and partly in the
abdomen, above the alimentary canal (14).

Each is a dark green mass, on the exterior of which minute rounded eminences (indications of the contained ova) can be seen. Near their anterior ends they lie in contact in the middle line, and for a short distance their substance is continuous.

ϵ. An oviduct arises from each ovary a little in front of its middle, and passes directly to the genital opening of its own side (1. *c.* γ.).

14. Alimentary organs.

a. Remove the dorsal part of the carapace in front of the cervical suture, and there will then be laid bare, in front of the position of the heart, a large sac—*the stomach;* pass a probe into it along the *gullet*, through the mouth-opening which lies between the mandibles.

b. Trace back the tubular *intestine* from the stomach to the anus. It dilates near the latter in the lobster. In the crayfish it presents a small cœcal diverticulum close to the stomach, and in the lobster one near the anus.

c. Examine the *liver*.

  a. It is an elongated soft pale-yellow mass lying in each side of the cephalo-thorax, and opening by a duct on each side at the point where the intestine joins the stomach.

  β. Tease out a bit of the liver in water; it is made up of branched cœcal tubes, which when examined microscopically are seen to be lined by a layer of cells (*epithelium*).

    d. Carefully remove the alimentary canal, cutting the gullet through close to the stomach.

        a. Open the latter under water and make out in it the constriction which divides it into an anterior (*cardiac*) and a posterior (*pyloric*) portion.

        β. The supporting bars and the hairs in the stomach, and the calcifications of its lining membrane.

15. Now trace the sternal artery (removing the alimentary canal and the genital organs), until it enters a passage (*sternal canal*) formed by ingrowths of the exoskeleton near the ventral surface of the animal. Just before entering this the sternal artery gives off the *inferior abdominal branch*, which runs back along the middle line of the abdomen immediately inside the sterna of the somites. Trace this branch back removing the muscles which cover it. By this proceeding the abdominal part of the nervous chain will be exposed. It lies immediately above the blood-vessel and is not to be injured. .

16. **The nervous system.**

        a. Find the *supraœsophageal ganglion* in front of the gullet.

        β. The *circumœsophageal commissures* passing back from it.

        γ. Follow back these commissures, cutting away the hard parts (forming the roof of the sternal canal) which come in the way; they lead to a chain of *six ganglia*, lying along the floor of the cephalothorax, and united by double cords (*commissures*).   Lying in the sternal canal beneath the ganglia may be seen the sternal artery (15).

   8.   Follow back the single cord proceeding from the last thoracic ganglion to the abdomen, removing any muscles which come in the way: it will lead to a chain of six ganglia, one for each abdominal segment, united by single cords.

17. **The green gland.**   A soft greenish mass lying on each side in the extreme front part of the cephalo-thoracic cavity: pass a fine bristle into it from the opening of its duct on the basal joint of the endopodite of the antenna (21. *f.*).

18. Tease out a bit of muscle in water and examine it microscopically: note its structure; it is made up of fibres, marked by regularly alternating transverse lighter and darker bands.

19. Tease out a bit of perfectly fresh nerve-cord in water and stain with magenta or hæmatoxylin.

   a.   Composed of slender fibres of varying size, each consisting of a structureless outer wall, on which are nuclei at intervals, surrounding a clear or, sometimes, finely granular or obscurely fibrillated central axis.

20. Tease out in water a ganglion which has been treated with osmic acid.

   a.   Composed of large oval branched cells, each consisting of a granular mass in which lies a clear round nucleus, containing a nucleolus.

21. **The appendages.** Beginning with the sixth abdominal segment, remove with forceps the appendages of the body and arrange them in order on a piece of cardboard. The abdominal appendages have been already described; note the following points in the remainder, working from behind forwards.

   a.   The four posterior thoracic appendages (*ambulatory appendages*).

      α.   The most posterior: elongated and seven-jointed, the joints working in different planes so that the limb as a whole can move in any direction: the joints have the following names; the proximal, short and thick, *coxopodite;* the next, small and conical, *basipodite;* next, cylindrical and marked by an annular constriction, *ischiopodite;* the next, longer, *meropodite;* then successively, the *carpopodite, propodite,* and *dactylopodite.* Probably the coxo- and basipodite together represent the protopodite of the abdominal appendages : the remaining joints the endopodite: the exo- and epipodite are wanting.

      β.   The next ambulatory leg: generally similar to the preceding, but possessing, attached to the coxopodite, a long membranous flattened appendage (*epipodite*) which ascends into the gill-chamber: it bears a gill.

      γ.   The next anterior ambulatory leg: differing from the last only in having its propodite prolonged so as to be opposable to the dactylopodite and form a pair of forceps (*chelæ*).

δ.   The most anterior ambulatory leg: resembling γ. closely and, like it, bearing a gill.

b.   *The great chelæ:* much larger and more powerful than the last appendage: but resembling it in structure, except that its ischio-podite and basi-podite are ankylosed together; it carries a gill.

c.   *The three maxillipedes.*

a.   The most posterior: its short thick basal two-jointed (*protopodite*): the three prolongations articulated to it; the external (*epipodite*) a curved elongated lamina lying in the branchial chamber and bearing a gill; the middle one (*exopodite*) long, slender and many-jointed; the internal one (*endopodite*) several-jointed and much resembling one of the ambulatory limbs.

β.   The middle maxillipede: much like a. but with the two joints of the protopodite fused together and with a less stout endopodite.

γ.   The anterior maxillipede; protopodite, exopodite and epipodite all present, but smaller than those of β. and the epipodite bearing no gill; the endopodite flattened and foliaceous.

The ambulatory limbs, great chelæ, and maxillipedes together constitute the appendages of the thorax; we now come to those of the head proper.

d.   *The two maxillæ.*

a.   The posterior: its protopodite and endopodite essentially like those of the anterior maxillipede; the epipodite and exopodite united and forming a wide oval plate (*scaphognathite*) which lies at the anterior end of the gill-chamber (11. γ.).

β.   Anterior maxilla: epipodite and exopodite undeveloped: the endopodite foliaceous.

ε.   *The mandible.* Its strong toothed basal joint (*protopodite*) bearing a small appendage (*the palp*) which represents the endopodite; the epipodite and exopodite unrepresented.

f.   *The antenna.* Its two-jointed basal portion (*protopodite*) bearing a flattened plate (the rudimentary *exopodite*) and a long multiarticulate filament (*the endopodite*): the opening of the green gland (17) on the oral side of the basal joint of the protopodite.

g.   *The antennula.* Its large trigonal basal joint (*protopodite*), bearing a pair of jointed filaments (*endopodite and exopodite*): the opening of the auditory organ (24) in the midst of a minute hairy tuft on the basal joint.

h.   *The ophthalmites or eyestalks.* Short two-jointed appendages representing only the basipodite.

22. Now work back over the 20 pairs of appendages and compare each with the third maxillipede: all may be supposed to be derived from it by suppression, coalescence or special change of form; it is what is called a *typical appendage.*

23. Structure of the Eye.

a.   Take the eye of a lobster which has lain four or five days in 0·5 per cent. solution of chromic acid and then twenty-four hours or more in alcohol. Examine its surface with one inch obj. with reflected light. It will be seen to be marked out into a great number of minute square areas or *facets*, each of which shews faint signs of furrows crossing it diagonally from corner to corner.

b. Imbed the eye and cut a number of sections from it perpendicular to its surface: mount in glycerine and examine with one inch objective.

    a. If the section has passed through the middle of the eye it will be seen to present a central mass (*optic ganglion*) from which a number of lines appear to radiate to the facets on the surface. These radiating lines (which are obscured here and there by concentric pigmented layers) are indications of the *striated spindles, connective rods* and *crystalline cones.*

c. Examine your thinnest section with a high power, or tease out one of your thicker ones in glycerine. Beginning at the exterior make out successively—

    a. The *cornea*, answering to one of the superficial facets. Its flat outer and slightly convex inner surface. Immediately beneath the cornea there will be seen (in good specimens) a slightly granular layer.

    β. The *crystalline cone*, an angular transparent body which is usually obscured by pigment. If this is the case, another section must be mounted in dilute caustic potash, which removes the pigment.

γ. Behind the crystalline cone comes *the connective rod.* It is widest in front where it joins the cone, but narrows posteriorly where it is continuous with the *striated spindle.* If fresh eyes be treated with osmic acid and then teased out, each of these rods can be split up into four fibres.

δ. The *striated body* is fusiform and presents well-marked transverse striations. Besides these coarse striations, however, much finer ones can be seen by careful examination with a high power. The outer ends of these spindles correspond in position to the second of the pigmented layers seen with the low power (*b. a.*): they are best seen in specimens treated with dilute caustic potash.

ε. Beneath the striated spindles is a perforated membrane through which the spindles pass to become continuous with the optic ganglion. From their ends pass nerve-fibres which run inwards in a converging manner and among which nerve-cells are here and there scattered. Within the ganglion are several concentric pigmented bands.

ζ. If the section has passed back along the optic nerve two obliquely placed lenticular masses will be seen among its fibres.

η. Passing back from the cornea to the optic ganglion is a membrane investing each cone, rod, and spindle. It is on this that most of the pigment lies which causes the two outer dark bands. Over the rods the pigment is wanting and there the membrane is seen to possess oval nuclei.

24. The Auditory organ.

This lies in the basal joint of the antennule and is best examined in the lobster. The upper surface of this basal joint is flat posteriorly and joins in front at an angle a rounded anterior portion. It bears several tufts of hairs: one of these is very small and lies at the inner side of the flattened surface, just at the angle where it meets the rounded part; among these hairs is the opening into the auditory 'sac, through which a bristle can easily be passed.

a.   Take a fresh antennule from a lobster and cut away the under surface of its basal joint. A chitinous transparent sac will readily be found in it, among the muscles &c.; this is the auditory sac and is about $\frac{1}{8}$ of an inch long. Carefully dissect it out.

b.   If this sac be held up to the light a little patch of gritty matter will be seen on its under surface near the aperture to the exterior. Behind this can be seen a curved opaque line; behind this, and concentric with it, a shorter brownish streak. Cut out carefully the part of the sac which bears these streaks: mount in sea-water or sodic chloride solution and examine with one inch objective.

a.   The white line will be seen to answer to a ridge on the apex of which is a row of large hairs, and both on the brown patch and on the opposite side of the main row will be seen scattered groups of smaller hairs.

c. Examine with ⅛ obj.

α. Each of the hairs seen with the lower power is
    now seen to be covered over its whole surface
    with innumerable very fine secondary hairs;
    these are shortest near the base of the primary
    hair. Towards its base each of the primary
    hairs is constricted and then dilates into a
    bulbous enlargement which is fixed to the
    wall of the sac.

β. The brown patch is seen to owe its colour to
    a single layer of polygonal epithelial cells
    containing pigment granules.

γ. By focussing through this epithelial layer a
    number of parallel slightly granular bands is
    seen passing up, one to the base of each hair
    in the main row on the top of the ridge. At
    the base of the hair to which it runs, each
    band is constricted and, entering the bulbous
    enlargement of the hair, joins a small hemi-
    spherical swelling within it.

δ. If a fresh auditory sac be put in 1 per cent.
    solution of osmic acid for half an hour, and
    then laid for twenty-four hours in distilled
    water and examined, each of the granular
    bands mentioned above is seen to consist of
    a bundle of fine fibres which swell out into
    fusiform enlargements at intervals.

ε. A great part of the whole interior of the audi-
    tory sac of the lobster is covered with very
    fine hairs which can only be seen with a high
    power. Epithelium is absent except the pig-
    mented patch above mentioned.

d. The auditory sac in the crayfish is very similar to that in the lobster, and may be examined in a similar way. It is however not so good, both on account of its smaller size and because the auditory hairs, although longer, are collected in a close tuft, which makes it more difficult to see the manner of their insertion.

XIII.

THE FROG (*Rana temporaria and Rana esculenta*).

THE only species of Frog indigenous in Britain is that termed the 'common' or 'Grass Frog.' (*Rana temporaria*), while, on the Continent, there is, in addition to this, another no less abundant species, the hind-limbs of which are considered a delicacy, whence it has received the name of the 'Edible Frog' (*Rana esculenta*). Unless the contrary be expressly stated, the description here given applies to both species. The Edible Frog is usually larger than the other, and is therefore more convenient for most anatomical and physiological purposes.

In the body of the Frog the head and trunk are readily distinguishable; but there is no tail and no neck, the contours of the head passing gradually into those of the body, and the fore-limbs being situated immediately behind the head. There are two pairs of limbs, one anterior and one posterior. The whole body is invested by a smooth moist integument, on which neither hairs, scales, nor other forms of *exoskeleton* are visible; but hard parts, which constitute the *endoskeleton*, may readily be felt through the integument in the head, trunk and limbs.

The yellowish ground-colour of the skin is diversified by patches of a more or less intense black, brown, greenish, or reddish-yellow colour, and, in the Grass Frog, there is a large, deep brown or black patch on each side of the head, behind the eyes, which is very characteristic of the species. The coloration of different frogs of the same species differs widely; and the same frog will be found to change its colour, becoming dark in a dark place, and light if exposed to the light.

The body of the Frog presents only two median apertures, the wide mouth and the small cloacal aperture. The latter is situated at the posterior end of the body, but rather on its upper side than at its actual termination. It is commonly termed the *anus*, but it must be recollected that it does not exactly correspond with the aperture so termed in the Mammalia.

The two nostrils, or *external nares*, are seen at some distance from one another upon the dorsal aspect of the head, between the eyes and its anterior contour. The *eyes* are large and projecting, with well-developed lids, which shut over them when they are retracted; and, behind the eye, on each side of the head, there is a broad circular area of integument, somewhat different in colour and texture from that which surrounds it; this is the outer layer of the membrane of the *tympanum*, or drum of the ear.

The fore-legs are very much shorter than the hind-legs. Each fore-limb is divided into a *brachium, antebrachium* and *manus*, which correspond with the arm, fore-arm and hand in Man. The manus possesses four visible digits which answer to the second, third, fourth, and fifth fingers in Man. There is no web between the digits of the manus.

The hind-legs are similarly marked out into three divisions, *femur, crus,* and *pes*, of which the femur answers to the thigh, the crus to the leg, and the pes to the foot, in Man. The pes is remarkable not only for its great relative size as a whole, but for the elongation of the region which answers to the tarsus in Man. It will be observed, however, that there is no projecting heel. There are five long and slender digits, which correspond with the five toes in Man, and are united together by thin extensions of the integument constituting the web. The innermost and shortest answers to the *hallux*, or great toe, in Man.

At the base of the hallux, the integument of the sole presents a small horny prominence, and sometimes there is a similar but smaller elevation on the outer side of the foot: but there are no nails upon the ends of any of the digits of either the pes or the manus. Thickenings, or *callosities*, of the integument, however, occur beneath the joints of the digits, both in the pes and the manus.

During the breeding season, the integument on the palmar surface of the innermost digit of the manus, in the male, becomes converted into a rough and swollen cushion, which, in the Grass Frog, acquires a dark-brown or black colour.

The Frog, when at rest, habitually assumes a sitting posture much like that of a dog or cat. Under these circumstances the back appears humped, the posterior half being inclined at a sharp angle with the anterior half. The vertebral column, however, will be found to be straight, and the apparent hump-back arises, not from any bend in the vertebral column, but from the manner in which the long iliac bones are set on to the sacrum.

The walk of the Frog is slow and awkward, but it leaps with great force, by the sudden extension of the hind-limbs, and it is an admirable swimmer.

In a living Frog, the nostrils will be seen to be alternately opened and shut, while the integument covering the under side of the throat is swollen out and flattened. The alternate pumping in and expulsion of the air needed for the Frog's respiration is connected with these movements.

The upper eyelid of the Frog is large and covered with ordinary pigmented integument, and it has very little mobility. What performs the function of the lower eyelid in Man, is a fold of the integument of which very little is pigmented and which is, for the most part, semi-transparent, so as to resemble the nictitating membrane of a bird rather than an ordinary lower lid. If the surface of the cornea be touched, the eyeball is drawn inwards under the upper lid, which descends a little, at the same time as the lower lid ascends over the ball, to meet the upper lid and close the eye.

As is well known, Frogs emit a peculiar croaking sound, their vocal powers being more especially manifested in the breeding season, when they collect together at the surface of ponds, pools and sluggish streams, in great numbers. At this season, which commences in the early spring for the Grass Frog, but much later on in the year for the Edible Frog, the male seeks the female and, clasping her body tightly with his fore-limbs, remains in this position for days or even weeks, until her ova are discharged, when he fecundates them by a simultaneous out-pouring of the seminal fluid. Shortly after the eggs pass into the water, the thin layer of viscid albumen, secreted by the oviduct, with which each egg is surrounded, swells up by imbibition and, with that which surrounds the others, it gives rise to a gelatinous mass in which the eggs remain imbedded during the early . stages of their development.

The development of the eggs is closely dependent upon temperature, being greatly accelerated by warmth and retarded by cold. The process of yelk-division, which commences within a few hours of impregnation, is readily observed when the eggs are examined as opaque objects under a low power of the microscope.

While still within the egg the embryo assumes the form

of a minute fish, devoid of limbs and with only rudiments of gills, but provided with two adhesive discs on the ventral side of the head behind the mouth.

After leaving the egg, the young acquires three pairs of *external branchiæ* having the form of branched filaments, attached to the sides of the hinder part of the head. Narrow clefts in the skin at the roots of the branchiæ lead into the back of the throat. Water taken in at the mouth passes out by these *branchial clefts*. The animal crops the aquatic plants on which it lives, by means of the horny plates with which its jaws are provided.

In the *Tadpole*, as the larval Frog is called, the intestine, which is relatively longer than in the adult, is coiled up like a watch-spring in the abdominal cavity. A membranous lip, the surface of which is beset with numerous horny papillæ, surrounds the mouth, and the muscular tail acquires a large relative size. The eyes, the nasal and auditory organs become distinct, but no limbs are at first visible.

A fold of the integument in the hyoidean region, called the *opercular membrane*, now grows back over the external gills and unites with the integument covering the abdomen, leaving only a small aperture on the left side, through which the ends of the external gills of that side may, for some time, be seen to protrude. The external gills atrophy and are succeeded functionally by short processes developed from the opposing faces of the branchial clefts—the *internal branchiæ*. The rudiments of the limbs appear, rapidly elongate and take on their characteristic shape, the hind pair only being at first visible on account of the anterior pair being hidden under the opercular membrane. The lungs are developed and, for a time, the tadpole breathes both by them and by its internal gills.

As the legs grow the tail shortens and, at last, is represented merely by the pointed end of the body; the gape elongates until the angle of the mouth lies behind the eye, instead of a long way in front of it, as in the tadpole; the labial membrane and the horny armature of the mouth disappear, while teeth are developed in the upper jaw and on the vomers; the intestine becomes less and less coiled as, not growing at the same rate as the body, it becomes relatively shorter; and the animal gradually changes its diet from vegetable to animal matters—the perfect Frog being insectivorous.

The two species, *Rana temporaria* and *Rana esculenta*, are distinguishable by the following external characters. In *Rana temporaria*, the interspace between the eyes is flat or slightly convex, and its breadth is usually greater than, or at least equal to, that of one of the upper eyelids. The diameter of the tympanic membrane is less than that of the eye, often much less. The horny elevation on the outer side of the pes is small or absent, and that on the inner is flattened and has a rounded margin. A patch of dark colour extends from the eye backwards over the tympanic membrane. The males have the cushion on the radial side of the manus black, and they are devoid of vocal sacs.

In *Rana esculenta*, on the other hand, the interspace between the eyes is usually concave and narrower than the breadth of one of the eyelids. The diameter of the tympanic membrane is as great as that of the eye. The horny elevation on the inner side of the pes is elongated, compressed and brought to a blunt edge, so as almost to resemble a spur, and a small outer elevation is constantly present. There is no patch of colour at the sides of the head, such as exists in *Rana temporaria*, and the cushion of the inner digit in the male is not black. The males have a large pouch on each side of the head, behind the angle of the jaw, communicating with the cavity of the mouth, and, when they croak, these pouches becoming dilated assûme the form of spherical sacs.

Having thus become acquainted with the general character and life-history of the Frog, and with those features of its organization which are visible to the naked eye and without dissection, its structure may next be studied in detail.

If the abdomen be laid open, it will be found to enclose a cavity in which some of the most important viscera—the stomach and intestine, the liver, the pancreas, the spleen, the lungs, the kidneys and urinary bladder, and the reproductive organs—are contained. As this cavity answers to those of the pleuræ and of the peritoneum in the higher animals, it is termed the *pleuroperitoneal cavity;* and the soft smooth membrane which lines it and covers the contained viscera is the *pleuroperitoneal membrane.*

The vertebral column traverses the middle of the roof of this cavity, and the layer of pleuroperitoneal membrane which lines each lateral wall of the cavity, passes downwards on each side of the vertebral column and joins its fellow in the middle line to form a thin sheet, the *mesentery*, which suspends the intestine. In the triangular interval left between these two layers before they unite, a wide canal—the *subvertebral lymph sinus*—the dorsal aorta, and the chain of sympathetic ganglia, are situated.

The dorsal moiety of the anterior end of the pleuroperitoneal cavity is occupied by the gullet, which places the mouth in communication with the stomach. Beneath the gullet the peritoneal cavity is separated only by a thin partition from a chamber, the *pericardium*, which contains the heart. The posterior face of the partition is constituted by the peritoneum, its anterior face by a membrane of similar character, the *pericardial membrane*, which lines the pericardium and is reflected on to the heart, in the same way as the peritoneum lines the peritoneal cavity and is reflected on to the intestine.

A layer of the muscular fibres which enter into the walls of the abdomen is continued inwards at the anterior boundary of the pleuroperitoneal cavity and is attached to the sides of the œsophagus and to those of the pericardium, thus constituting a rudimentary *diaphragm;* which, it will be observed, is situated in front of the lungs, and not behind them, as in the higher animals.

Thus, in the trunk, on the ventral side of the vertebral column, the body presents two cavities, one large posterior pleuroperitoneal cavity, and one small, anterior to the foregoing, the pericardial cavity, and neither of these communicates directly with the exterior, though in the female there is an indirect communication by the oviducts.

On the ventral side of the head, the very wide mouth opens into a spacious buccal cavity, the roof of which is hard and firm, while the floor is soft and flexible, except so far as the middle of it is occupied by a broad, flat, for the most part gristly plate, the *body* of the *hyoid* bone.   Within the lips the upper jaw is beset with numerous sharp small teeth, and two clusters of similar teeth are to be seen in the fore part of the roof of the mouth; the latter, being attached to the bones termed the *vomers*, are the *vomerine* teeth, while the former, attached to the *premaxillæ* and *maxillæ*, are *maxillary* teeth.   The lower jaw or *mandible* bears no teeth.

At the sides of the clusters of vomerine teeth are the apertures termed *posterior nares*, by which the nasal chambers communicate with the mouth.   At the sides of the back part of the throat two wide passages, the *Eustachian recesses*, lead into the *tympanic* cavities, which are closed externally by the tympanic membranes. In the male *Rana esculenta* the small apertures of the vocal sacs are seen on the inner side of each ramus of the jaw, close to the angle of the gape and nearly opposite the Eustachian recesses. In the middle of the back of the throat is the opening of the œsophagus, closed by the approximation of its sides except during deglutition, while in the median line of the hinder part of its floor lies a longitudinal slit, the *glottis*. A fleshy tongue, bifurcated and free at its posterior end, is attached anteriorly to the middle part of the lower jaw. In a state of rest, therefore, it lies on the floor of the mouth with its free end turned backwards, and one point on each side of the glottis.

The gullet, after traversing the diaphragm, passes into the elongated stomach. At its posterior end this narrows and joins the slender *small intestine.* Though short, this is too long relatively to the length of the abdominal cavity to lie straight in it. It is, therefore, thrown into sundry folds which are suspended to the dorsal wall of that cavity in the manner before described. Finally, the small intestine enters the suddenly dilated short *large intestine*, and this opens into a chamber with muscular walls, the *cloaca*, the external aperture of which has been already mentioned.

Thus the alimentary canal is a tube which traverses the body from the oral to the anal apertures; and the heart, enclosed in the pericardium, is situated in the middle line on the ventral side of the alimentary canal.

Separated from the pleuroperitoneal and oral cavities by the bodies of the vertebræ and the hard roof of the oral chamber which continues the direction of these forwards, is an elongated cavity, widest in the head but becoming very narrow posteriorly, which is closed on all sides by the bony and other elements of the head and spinal column. This is the *neural* cavity and contains the brain and spinal cord, which together constitute the *cerebro-spinal nervous axis*. The neural cavity is lined by a serous membrane resembling the peritoneum and the pericardium, and this *arachnoid membrane* is reflected on to and covers the contained cerebro-spinal axis, so that the latter is related to it as the heart is to the pericardial membrane.

The *cerebro-spinal nerves* which are given off from the brain and spinal cord pass to their destination through the boundary walls of the neural cavity.

A transverse section of the head in the region of the eyes will shew, in the middle line, a dorsal cavity in which the anterior part of the cerebro-spinal axis, the *brain*, is contained, separated by the solid floor of the skull from a ventral cavity, the mouth.

A transverse section of the abdomen will shew a dorsal cavity containing the posterior part of the cerebro-spinal axis, the *spinal cord*, separated by the solid floor of the vertebral column from a ventral cavity containing the alimentary canal and continuous with that of the mouth. But the backward continuation of the alimentary canal is embraced by the large pleuroperitoneal chamber, of which there is no indication in the head.

On comparing the transverse section of the abdomen of the Frog with a transverse section of the middle of the body of the Lobster, it will be seen that while the chief nervous centre is on one side of the alimentary canal and the heart on the opposite side in both cases, there is no solid and complete partition between the nervous centre of the Lobster and the alimentary canal. Moreover, the face of the body on which the nervous centre lies, is that on which the Lobster naturally rests, while in the Frog it is the reverse. The limbs are turned towards the neural side in the Lobster and away from it in the Frog, and the like difference obtains between all *Vertebrata* and all *Arthropoda*.

Using the term *skeleton*, in its broadest sense, for the framework which protects, supports and connects the various parts of the organism, it consists in the Frog of four kinds of tissue; the Horny, the Osseous, the Cartilaginous and the Connective. Moreover, the hard parts are either developed in the integument, constituting an *exoskeleton*, or they are deeper seated and belong to the *endoskeleton*.

Leaving aside a question that may arise as to the nature of some of the cranial bones, the exoskeleton in the Frog is almost absent, being represented only by the horny coating of the calcar.

The endoskeleton, on the contrary, is well developed and, as in all the higher Vertebrata, may be distinguished into an axial and an appendicular portion.

The *axial endoskeleton* consists of the notochord, the spinal column and the skull.

The *appendicular endoskeleton* occurs in the limbs and in the pectoral and pelvic arches to which they are attached.

In the order of development, the endoskeleton is at first represented by the notochord alone; secondly, nascent connective tissue and cartilage are superadded to the notochord; thirdly, these acquire their special characters; fourthly, they become replaced by bone.

The process of conversion or replacement indicated under the last head is very incomplete, even in the adult Frog, in which remains of the notochord are to be found in the centres of the vertebræ; and the cartilage, of which the greater part of the skeleton at one period of larval existence was composed, to a great extent persists.

Such cartilage is found forming the free surfaces of the

bodies of the vertebræ, the extremities of the caudal style (*urostyle*) and the ends of the transverse processes; and it enters largely into the sternum.   In the skull, the parasphenoid[1], vomers, parieto-frontals, nasals, premaxillæ, maxillæ, jugals, squamosals, and the bony elements of the mandible may be removed by maceration, leaving behind the primitive cartilaginous skull, or *Chondro-cranium*, altered only so far as parts of it have been replaced by bone.

It furnishes a floor, side walls and roof to the brain-case, interrupted only by a large space (called a *fontanelle*) covered in by membrane, which lies in the inter-orbital region under the parieto-frontals, and by the foramina for the exit of the cranial nerves.   It consists entirely of cartilage, except where the exoccipitals, the pro-otics, and the sphenethmoid invade its substance.   In front of the septum of the anterior cavity of the sphenethmoid, it is continued forward between the two nasal sacs, as the cartilaginous partition between the nasal cavities (*septum narium*), from which are given off, dorsally and ventrally, transverse alæ of cartilage which furnish a roof and a floor, respectively, to the nasal chambers. Of these, the floor is the wider.   The dorsal and ventral alæ pass into one another where the chondro-cranium ends anteriorly and give rise to a truncated terminal face, which is wide from side to side, narrow from above downwards, and convex in the latter direction.   The lateral angles of this truncated face are produced outwards and forwards into two flattened *præ-nasal processes;* these widen externally and end by free edges which support the adjacent portions of the premaxillæ and maxillæ.   From the ventral face, just behind the truncated anterior end of the chondro-cranium, spring

---

[1] See Laboratory work, D, c, for the structure of the bony skull (*osteocranium*), which should be thoroughly understood before any attempt is made to study the cartilaginous skull or *chondro-cranium*.

two slender cartilages, the *rhinal processes*. Each of these
inclines towards the middle line and ends against the middle
of the posterior face of the ascending process of the pre-
maxilla by a vertically elongated extremity. An oval nodule
of cartilage is attached to the posterior face of the dorsal
end of the ascending process of the premaxilla, and serves
to connect it with the rhinal process. On the dorsal face of
the chondro-cranium, just above the point of attachment of
the rhinal processes, the external nasal apertures are situated,
and the outer and posterior margins of each of these apertures are surrounded and supported by a curious curved
process of the cartilaginous ala—the *alinasal process*. Where
the sphenoidal and the ethmoidal portions of the sphenethmoid meet, a stout, transverse, partly osseous and partly
cartilaginous bar is given off, which is perforated at its origin
by the canal for the orbito-nasal nerve. It then narrows,
but becoming flattened from above downwards, rapidly
widens again, and its axe-head-like extremity abuts against
the inner face of the maxilla. The anterior angle of the
axe-head is free; the posterior angle is continued back into
a slender cartilaginous *pterygoid* rod which bifurcates posteriorly. The outer division passes into the ventral crus of
the suspensorium. The inner division is the *pedicle of the
suspensorium;* it articulates by a joint with the anterior face
of the broad lateral process of the hinder part of the chondro-cranium, which contains the auditory labyrinth and is
termed the *periotic capsule*. The *Suspensorium* is a rod of
cartilage, which lies between the squamosal and the pterygoid bones and, at its distal end, articulates with *Meckel's
cartilage* which forms the core of the ramus of the mandible.
At its dorsal end it divides into two divergent processes or
crura, of which the ventral *crus* has already been said to be
continuous with the pterygoid. The dorsal *crus*, on the

other hand, passes upwards and, curving backwards, becomes attached to the dorsal part of the outer face of the periotic capsule.

Meckel's cartilage, articulated to the free end of the suspensorium, is unossified throughout the greater part of its extent, no osseous *articulare* being developed; but, at its symphysial end, each cartilage becomes ossified, and forms the *mento-Meckelian* element of the mandible.

The slender, cartilaginous band (*cornu* of the *hyoid*) by which the body of the hyoid is attached to the skull, is connected with the periotic cartilage immediately in front of and below the fenestra ovalis.

The pectoral and pelvic arches (see Laboratory work D. e. g.) are, in the young state, undivided cartilages on each side, and the development of bone in and upon them does not really destroy this continuity, the cartilage persisting at the ends of the bones and between them, in the glenoidal and acetabular cavities.

In like manner, the bones of the limbs consist originally of merely cartilaginous models of the perfect bone; but, as development proceeds, the middle of the cartilaginous model commonly becomes invested by a sheath of true bone, while a calcareous deposit takes place in the cartilage close to its growing extremities. As the bone grows, the superadded sheath invades the middle of the cartilage and more or less replaces it; while the terminal portions of cartilage continue to grow and enlarge and the calcareous deposit within them increases, without however reaching their surfaces. Thus one of the larger adult limb-bones, the femur, consists of a median shaft of perfect bone, and of two terminal cones of cartilage, containing calcified *epiphyses*, enclosed within and more or less overlapping the hollow ends of the shaft.

The general disposition of the parts which are seen in the mouth has already been described.

Teeth are found attached only to the premaxillæ, maxillæ and vomers. They are small, with recurved and pointed crowns. New teeth are constantly being developed in the gum to replace those which are worn out or broken away. And as they attain their full size these teeth become ankylosed to processes of the subjacent bone.

The gullet passes without change of diameter into the stomach, which lies on the left side of the abdominal cavity and is nearly as long as it. The stomach narrows posteriorly and the almost tubular pyloric portion bends round sharply and passes into the duodenum. A slight constriction marks the pylorus. The duodenum runs forwards parallel with the stomach, so that with the latter it forms a sort of loop. At its anterior end it passes into the rest of the small intestine (*ileum*), which is coiled up into a sort of packet and lies on the right side of the abdominal cavity, being held in its place by a mesenteric fold of the peritoneum. From the comparatively narrow neck of the packet, the small intestine proceeds backwards in the middle line and opens into the anterior end of the dilated large intestine or *colon* and *rectum*.

The inner wall of the stomach is raised up into a number of strong longitudinal folds which project into its cavity and give it a stellate appearance in transverse section. Much more delicate continuations of these folds are continued into the small intestine and are there joined by transverse folds.

The opening of the ileum into the colon is valvular, its edges projecting backwards into the cavity of the colon. On the dorsal aspect, this presents a slight forward dilatation, which may be regarded as a rudiment of a cœcum.

The liver is very large, and is divided into two lobes united by a mere bridge, dorsally and anteriorly. The left lobe is further subdivided into two. The gall bladder is attached to the posterior and dorsal face of the right lobe. The biliary duct opens into the duodenum, at some distance behind the pylorus, and its termination is embraced by the base of the slender pancreas.

The rounded spleen lies in the mesentery, projecting more to the left than to the right side, just above the point at which the duodenum passes into the ileum.

The apparatus of circulation in the Frog consists of the blood and lymph vessels and their contents.

The lymph is a colourless fluid containing colourless nucleated corpuscles which exhibit amœboid movements: it is contained partly in large spaces immediately beneath the integument; in the pleuroperitoneal cavity and pro-bably in the other serous cavities; and, partly, in capillaries and larger trunks which are interlaced with and accompany the blood-vessels. The largest of the trunks is the great sub-vertebral lymph-sinus, which lies between the layers of the root of the mesentery and communicates by small pores with the pleuroperitoneal cavity. There are four lymph-hearts.

The blood consists of a colourless plasma which contains colourless corpuscles, similar to those of the lymph, and in addition a great number of oval nucleated red corpuscles. It is contained in the blood-vessels, which consist of capil-laries, arteries and veins, the two latter being connected on the one side by the capillaries and, on the other, by the heart into which they open. The lymphatics and the blood vessels are brought into connexion with one another by communications between the anterior lymph-hearts and the innominate veins, and between the posterior lymph-hearts and the iliac veins.

The heart is connected with the walls of the pericardium, on which spots of pigment may be observed, by the vessels which enter and leave it, and by a slender band which passes from the dorsal face of the base of the ventricle to the posterior and dorsal wall of the pericardial chamber.

The heart consists of four readily distinguishable segments, (1) the *sinus venosus*, (2) the *atrium*, (3) the *ventricle*, and (4) the *truncus arteriosus*, disposed in such a manner that the sinus venosus, which is the hindermost division, lies in the middle line on the dorsal aspect of the heart: the atrium is also median and on the dorsal side, but is in front of the sinus venosus; the ventricle is median, ventral and posterior; and the truncus passes obliquely forwards from the right side of the ventricle and is ventral and anterior. The heart therefore may be compared to a tube divided by constriction into four portions and bent somewhat into the shape of an S.

The *sinus venosus* receives on each side, in front, a large vein, the *vena cava superior;* while behind the, usually single, *vena cava inferior* opens into it. It opens by a valvular aperture into the atrium. The latter shews no signs of division externally, but, internally, it is divided by a delicate partition, the *septum of the auricles*, into a smaller *left auricle* and a larger *right auricle*. The sinus venosus opens into the atrium, to the right of the septum and therefore into the right auricle. Into the left auricle, the *common pulmonary vein*, a small trunk formed by the junction of the veins from the right and left lungs, opens.

At its posterior end the atrium opens by the auriculoventricular aperture into the ventricle.

A small valve, prevented from flapping back by fine ten-

dinous cords, exists on each side of this aperture, and the
septum of the auricles is continued back upon the faces of
these valves and ends by a free edge between them, thus
dividing the auriculo-ventricular aperture itself into two
openings.

The walls of the sinus and of the atrium are very thin.
Those of the ventricle, on the other hand, are thick and
spongy, only a comparatively small, transversely elongated,
clear cavity being left at the anterior end or base of the
ventricle. At the right extremity of this is the aperture
which leads into the truncus arteriosus. Three semilunar
valves, which open from the ventricle into the truncus, surround this opening.

The walls of the *truncus arteriosus* are thick and muscular, though not nearly so thick as those of the ventricle.
At its anterior end it appears to divide into two trunks,
which diverge and immediately leave the pericardium to
pass on to the sides of the gullet. The elongated undivided part is the *pylangium*, the terminal part common to
the divergent trunks is the *synangium*. The former is
divided throughout its length by a sort of fold which is
attached to the dorsal wall while its opposite edge is free.
Three semilunar valves separate the pylangium from the
synangium, in which are the openings, posteriorly, of the
pulmonary arteries, anteriorly of the *carotid trunks;* while,
at the sides, the cavity of the synangium opens into those
of the right and left aortic arches. The apparently simple
branches into which the *truncus arteriosus* divides, are, in
fact, each made up of three separate trunks, the *pulmo-
cutaneous trunk* behind, the *aortic arch* in the middle and
the *carotid trunk* in front.

When the heart is in action, the sinus venosus, the atrium,
the ventricle and the truncus arteriosus contract in the

order in which they have just been named. Each contracts as a whole, so that the two auricles are emptied simultaneously. The blood from each is forced into the corresponding half of the spongy cavity of the ventricle, so that the right half of the ventricle contains venous blood and the left arterial blood. When the systole of the ventricle takes place, the blood which is first driven into the truncus arteriosus (the opening of which is, as has been seen, at the right end of the cavity) is therefore venous. It fills the conus arteriosus and, finding least resistance in the short and wide pulmonary vessels, passes along the left side of the median valve into them. But as they become distended and less resistance is offered elsewhere, the next portion of blood, consisting of the venous and arterial blood which have become mixed in the middle of the ventricle, passes on the right side of the longitudinal valve into the aortic arches. And, as the truncus becomes more and more listended, the longitudinal valve, flapping over, tends more and more completely to shut off the openings of the pulmonary arteries and to prevent any blood from flowing into them.

Finally, the last portion of blood from the ventricle, representing the completely arterialized blood of the left auricle which is the last to arrive at the opening of the truncus, passes into the carotid trunks and is distributed to the head.

The principal vessels of the Frog are disposed as follows :—

A. *Arteries.*

    1. The system of the *anterior aortic arch* (*carotid* trunk).

       a. Lingual artery—to the tongue.

    b.   *Carotid* artery—to the interior of the skull and the brain.

2. The system of the *middle aortic arch* (*aortic* trunks).

    a.   *Vertebral* and *subclavian*—to the vertebral column, and to the fore-limb. *Œsophageal* to the gullet.

    b.   *Cæliaco-mesenteric* (given off from the left arch, or from the dorsal aorta, at, or beyond, the junction of the two arches).

        α. *Cæliac* to stomach and liver.

        β. *Mesenteric* to intestine and spleen.

    c.   Branches of the dorsal aorta to the adrenal and renal organs, to the genital organs and to the muscles of the back.

    d.   The terminal branches of the dorsal aorta (*common iliac*); each of these gives off *hypogastric* arteries to the bladder and walls of the abdomen and is continued as the *femoral artery* into the leg.

3. The system of the *posterior aortic arch* (*pulmocutaneous* trunk).

    a.   *Pulmonary* artery to the lungs.

    b.   *Cutaneous* artery to the dorsal integument.

B. *Veins.*

1. The system of the *superior cava* formed on each side by the union of the *vena innominata*, the *subclavian* and the *external jugular.*

    a.   *Internal jugular* vein: leaves the skull by the jugular foramen, and brings back blood from the brain, spinal cord and anterior vertebral region.

b. *Subscapular:* returns the blood from the brachium and shoulder. These two veins (*a.* and *b.*) unite to form the *vena innominata.*

c. The *musculo-cutaneous vein,* receiving the blood of the surface of the head (except the mandibular and hyoidean regions) and that of the . back of the trunk—passes forwards between the internal and external oblique muscles of the abdomen.

d. The *brachial vein* receives blood from the antebrachium and manus.

These (*c.* and *d.*) unite to form the *subclavian* vein.

e. The veins of the mandibular region and those of the tongue unite into the *external jugular* vein.

2. The system of the *inferior cava,* formed by the union of the *renal, genital* and *hepatic* veins.

a. The *femoral* vein from the front of the leg, and—

b. The *sciatic* vein from the back of the leg, pour their blood into a trunk which lies in the lateral wall of the pelvis and may be termed the *pelvic* vein; the dorsal end of this becomes—

c. The *common iliac vein,* which passes to the outer edge of the kidney and is distributed to that organ, whence the blood is carried to the vena cava inferior by the *renal* veins.

d. The *dorso-lumbar* vein, which lies along the
transverse processes of the vertebræ and receives
blood from the walls of the abdomen and from
the interior of the spinal canal, opens into the
common iliac.

3. The system of the *anterior abdominal vein*, formed
by the union of the ventral ends of the pelvic veins
(2. *b.*). It receives blood from the urinary bladder
and the walls of the abdominal cavity, and at its
anterior end divides into two branches—a right and
a left. These branches go to the corresponding
lobes of the liver, the left receiving a large communicating branch from the gastric division of the *vena
portæ.*

4. The system of the *vena portæ* formed by the union
of two veins; one, *gastric*, which brings back the
blood from the stomach, the other, *lieno-intestinal*,
which returns that from the spleen and intestines.

[Hence the right lobe of the liver and part of the left
lobe are supplied with systemic venous blood, more
or less mixed with gastric venous blood, while only part
of the left lobe is supplied with intestinal venous blood.
Besides this venous blood, it must be recollected that
the liver receives arterial blood by the hepatic artery.]

5. The system of the *pulmonary vein*, formed by the
union of the veins of the right and left lungs.

In addition to the apparatus of the circulation of the
blood, the Frog possesses two pairs of *lymph-hearts*. These
are contractile muscular sacs, which are connected on the
one hand with the lymphatic vessels and on the other with
large veins in their neighbourhood; and which pump the

lymph contained in the wide lymphatic vessels and in the pleuro-peritoneal cavity of the Frog, into these veins.

The anterior lymph-hearts are situated close to the transverse processes of the third vertebra, below the edge of the scapula; the posterior pair lie one on each side of the uro-style, and their pulsations may be observed by carefully watching the integument in this region in a living Frog.

The *Thymus* gland is a small rounded body situated immediately behind the suspensorium, in a position corresponding to the dorsal ends of the obliterated branchial arches.

The *Thyroid* gland appears to be represented by two or more oval bodies, which are found attached to the lingual vessels and between the aortic and pulmo-cutaneous trunks.

The *Adrenal* glands are yellow bodies imbedded in the ventral face of the kidney.

The slit-like glottis of the Frog is formed by the apposition of two longitudinal folds of the mucous membrane of the mouth, each of which contains a cartilage of similar form. These cartilages are the *arytenoid* cartilages. They are articulated with an annular cartilage (*laryngo-tracheal*) which supports the wall of the very short chamber which represents the larynx and trachea. When the two folds of the glottis are divaricated, there are seen between them two membranous pouches, the free edges of which meet in the middle line, while anteriorly and posteriorly they pass into the mucous membrane which lines the faces of the longitudinal folds. These are the *vocal ligaments*, and the slit between them is what answers to the glottis in Man. It is by their vibration that the croak of the Frog is produced.

Laterally the laryngo-tracheal chamber opens into the

lung of each side. The lung is a transparent oval sac, somewhat pointed posteriorly, which lies at the side of the œsophagus in the dorsal region of the abdominal cavity. It is covered by a layer of the pleuroperitoneal membrane which represents the visceral layer of the pleura in the higher animals. The wall of the pulmonary sac is produced inwards so as to give rise to septa, which are much more prominent and more numerous in the anterior than in the posterior part of the lung and divide the periphery of the cavity into numerous air-cells, on the walls of which the ramifications of the pulmonary artery are distributed.

The lungs are elastic, the distended lung collapsing suddenly when it is pricked, and they contain abundant muscular fibres.

Inspiration is effected in the Frog by a buccal force-pump. The mouth being shut and the external nostrils open, the floor of the mouth is depressed, and the buccal cavity fills with air. The nostrils being then shut, the hyoid, and with it the floor of the mouth, is raised, the aperture of the gullet being at the same time closed. Thus the air is forced through the glottis and distends the lungs.

In ordinary expiration, the elasticity of the lungs and the pressure of the surrounding viscera probably suffice to expel the air; but this operation may be powerfully aided, firstly by the contraction of the intrinsic muscular fibres of the lungs; secondly, by the contraction of the muscles of the lateral and ventral regions of the abdominal wall; and, thirdly, by the contraction of those muscular fibres which enter into the diaphragm; as all these actions tend, either directly or indirectly, to diminish the capacity of the lungs.

It is essential to inspiration that the mouth should be shut, and it is said that frogs may be asphyxiated by keeping their mouths open.

In addition to its principal pulmonary apparatus of respiration, the Frog has a secondary respiratory apparatus in its moist and delicate skin. A considerable amount of venous blood is, in fact, constantly supplied to this organ by the large cutaneous branch of the pulmo-cutaneous artery. It has been experimentally ascertained that frogs in which the lungs have been extirpated will continue to live and respire for a considerable time, especially at a low temperature, by means of the skin.

The kidneys are elongated and flattened from side to side, and are kept in their places by the continuation of the peritoneum over their ventral faces. The ducts of the kidneys pass from about the junction of the middle and posterior thirds of the outer edge of each kidney and, approaching as they pass backwards, open by two small closely approximated slit-like apertures in the posterior wall of the cloaca.

The urinary bladder is a large bilobed sac, opening posteriorly, by a wide median aperture, into the anterior end of the cloaca, on the ventral side of the rectum.

The *testes* are spheroidal yellowish bodies situated in front of the kidneys and enveloped in peritoneum, a fold of which, forming a sort of testicular mesentery or *mesorchium*, passes into that which covers the ventral face of the kidney. The delicate *vasa efferentia* of the testes may be seen traversing this fold to enter the substance of the kidney. They communicate with the urinary tubules, and thus the duct of the kidney serves not only as the duct of the urinary excretion but as the *vas deferens*.

The spermatozoa of *Rana esculenta* have thick and cylindrical heads, while those of *Rana temporaria* are linear.

The *ovaria* are broad lamellar organs, very large and much folded and plaited in the breeding season. The in-

terior of each is hollow, and is divided into several chambers.
Innumerable ovisacs, containing dark-coloured ova, are scattered through the substance of the ovary and give rise to
projections upon the inner surface of the ovarian chamber
as they become fully developed.

The *oviducts* are long convoluted tubes situated on each
side of the dorsal wall of the abdominal cavity to which they
are connected by peritoneal folds ; each curves over the
outer face of the root of the lung.   Their anterior ends are
very slender, and terminate by open mouths at the sides of
the pericardium, between the attachment of the diaphragm
and the lobe of the liver.   The fold of peritoneum which
serves as a ligament, holding the lobe of the liver to the
diaphragm, œsophagus and posterior wall of the pericardium, in fact constitutes the outer lip of the oviducal aperture.   For the greater part of their length their walls are
thick and glandular, and swell up when placed in water.
Posteriorly, the oviducts dilate into capacious thin-walled
chambers and end, close together, by openings which are
situated in the dorsal wall of the cloaca immediately in front
of the apertures of the ureters.

Each ovum, when ripe, consists of a structureless vitelline
membrane, inclosing a vitellus, within which is a germinal
vesicle, containing several 'germinal spots.'   One half of
the vitellus is deeply coloured, the other pale.

The actions of the different parts of the organism of the
Frog are coordinated with one another and brought into
relation with the external world by means of the muscular
and nervous systems and the organs of sense.

The *muscles* consist partly of striped and partly of un-
striped fibres, the former being confined to the muscles of
the head, trunk and limbs and the heart, while the latter

are found in the viscera and vessels. An account of the disposition of the muscles in the hind-limb will be found in the Laboratory work.

The *nervous system* is conveniently divisible into two parts, the *cerebro-spinal* and the *sympathetic.* The cerebro-spinal nervous system again consists of the brain, or *encephalon*, with its nerves, and the spinal cord, or *myelon*, with its nerves.

The encephalon lies in the cranial cavity, which it nearly fills, and is divisible into the *hind-brain*, the *mid-brain* and the *fore-brain*, which last again comprises three divisions ; the thalamencephalon, the cerebral hemispheres, and the olfactory lobes.

The greater part of the hind-brain is formed by the *medulla oblongata*, which is the continuation of the myelon forwards and presents, on its dorsal aspect, a triangular cavity, the apex of which is directed backwards. It is roofed over by a thick and very vascular membrane (choroid plexus), the inner surface of which presents transverse folds on either side of a median longitudinal ridge. The cavity is the *fourth ventricle ;* it communicates behind with the central canal of the myelon, while, in front, it narrows into a passage which connects the fourth ventricle with the cavities anterior to it. The thick lateral ridges of nervous substance at the sides of the fourth ventricle, which represent the *restiform bodies*, pass, in front, into the outer extremities of a short broad tongue-shaped plate, convex ventrally and concave dorsally, which overhangs the anterior part of the fourth ventricle, and is the *cerebellum.*

In front of this, the dorsal moiety of the mid-brain is formed by two oval bodies, the long axes of which are directed inwards and backwards. These are the *optic lobes.* When laid open, each is seen to contain a cavity or ventricle with an opening on its inner face. These openings lead into a short passage, which communicates with the *iter a tertio ad quartum ventriculum*, as the canal which leads, through the mesencephalon, from the fourth to the third ventricle is termed. The floor of this canal is formed by the thick principal mass of the cerebro-spinal axis. It exhibits a median longitudinal depression or raphe, and in this region represents the *crura cerebri*.

In front of the mid-brain comes the hinder division of the fore-brain, or *thalamencephalon*, which is very distinct in the Frog and contains a median cavity, the *third ventricle*. On each side, the cavity of the third ventricle is bounded by a thick mass of nervous matter into which the crura cerebri pass. These are the *optic thalami*. Dorsally, the walls of the third ventricle are very thin and easily torn through, except behind, where there is a thick transverse band of nervous substance, the *posterior commissure*.

From the fore part of the roof of the third ventricle, a delicate process proceeds to the *pineal gland*—an ovate body lodged between the posterior parts of the cerebral hemispheres. The front part of the floor of the ventricle, on the other hand, is produced into a bilobed process directed backwards, which is the *infundibulum*. This is connected below with the *pituitary body*. In front of this is seen the commissure of the optic nerves.

Anteriorly, the third ventricle is bounded by the thick *lamina terminalis* which contains the *anterior commissure*. On each side, between this and the peduncle of the pineal gland, is a small aperture, the *foramen of Munro*, which leads into a cavity in the interior of the cerebral hemisphere —the *lateral ventricle*.

The hemispheres are elongated bodies, broader behind than in front, where they are marked off only by a slight constriction from the olfactory lobes. The outer wall of the ventricle, though relatively thick, presents nothing which can be called a distinct *corpus striatum*. The inner wall forms one or two convex projections into the ventricle.

In the bases of the olfactory lobes the forward continuation of the ventricular cavity is very narrow and the lobes become nerve-like cords, which leave the skull and spread out on the posterior faces of the olfactory sacs.

The inner faces of the hemispheres are quite free and separated by a cleft, the *great fissure*, but the inner faces of the commencements of the olfactory lobes are closely united together, giving rise to a kind of *corpus callosum*.

There are ten pairs of cranial nerves ordinarily so called, though it is to be recollected that the first and second pairs are proved, by their development, to be lobes of the brain.

1. *Olfactorii.*

   The olfactory lobes are what answer to the so-called olfactory nerves of the higher Vertebrata. They are distributed exclusively to the olfactory sacs.

2. *Optici.*

   These diverge from the base of the brain in front of the infundibulum. They are originally outgrowths of the thalamencephalon which secondarily become connected with the optic lobes.

Of the remaining cranial nerves five pairs leave the skull in front of the auditory capsules, while one pair enters those capsules and two pairs pass out behind the capsules.

The *Præauditory nerves* are the following.

3. *Motores oculorum*

   arise from the front part of the floor of the mid-brain and are distributed to all the muscles of the eye except the external rectus, the superior oblique and the retractor bulbi.

4.   *Pathetici*

arise from the floor of the mid-brain and pass out, on the dorsal aspect of the brain, between the cerebellum and the optic lobes.   They are distributed to the superior oblique muscles of the eye.

5.   *Trigemini*

take their origin in the front part of the floor of the hind-brain and, passing out at its sides, each dilates into a yellow enlargement—the *Gasserian ganglion*—which lies, in front of the auditory capsule, in the foramen of the pro-otic bone by which the nerve, after leaving the ganglion, passes out of the skull.

This ganglion is connected with the trunk of the sixth and seventh nerves and with the anterior end of the sympathetic, and some of the branches which appear to be given off from it really belong to the sixth and the seventh nerves.   Beyond the ganglion, the nerve divides into three main branches, the *orbito-nasal*, the *palatine* and the *maxillo-mandibular*.

i.   The *orbito-nasal* (usually termed the first division of the fifth nerve) is distributed :

a.   To the external rectus.

b.   To the retractor of the bulb.

(These branches (*a.* and *b.*) belong to the sixth nerve.)

c.   A branch which anastomoses with the fourth nerve.

d.   A branch to the Harderian gland.

e.   The principal trunk of the nerve passes through

the ant-orbital process of the skull into the nasal chamber and is finally distributed to the nasal mucous membrane and to the integument of the nose.

ii.   The *palatine* is distributed :

    a.   To the roof of the oral cavity.

    b.   Its main trunk runs forward between the mucous membrane of the roof of the mouth and the skull, pierces the vomer and ends in the mucous membrane of the anterior part of the palate.

    (This nerve is chiefly, if not wholly, derived from the seventh nerve.)

iii.   The *maxillo-mandibular* divides into two trunks, usually termed the second and third divisions of the fifth nerve.

    a.   *Maxillary*, passes outside the eye and is distributed to the integument of the upper jaw; an anastomotic branch unites this nerve with the palatine.

    b.   *Mandibular*, passes between the temporal and pterygoid muscles, below the jugal, over the articulation of the mandible and along the inner face of the latter, to the symphysis, giving off branches to the integument, muscles, teeth and tongue.

6.   *Abducentes*

arise from the floor of the hind-brain and leave the ventral surface of the medulla oblongata close to the middle line.   Each then unites so closely with the Gasserian ganglion and with the orbito-nasal division of the fifth as to appear to be only a subdivision of the latter (see 5. i. *a.* and *b.*).

7. *The Faciales*

take their origin from the floor of the hind-brain, behind the fifth and in common with the eighth; and, leaving the hind-brain, enter into close connexion with the Gasserian ganglion. Each then divides into two branches, an anterior and a posterior. The anterior passes into the palatine division of the fifth; the posterior passes between the dorsal and ventral crura of the suspensorium, enters the tympanic cavity, runs over the *columella auris* and then, as it leaves the tympanum, receives a very large branch from the glossopharyngeal. Finally it divides into two branches, anterior and posterior.

    a.   The former, which answers to the *chorda tympani* of the higher Vertebrata, runs along the inner face of the ramus of the mandible parallel with the mandibular branch of the fifth.

    b.   The posterior passes alongside the cornu of the hyoid and supplies its muscles.

8. The *Auditorii*

arise in common with the foregoing. Each divides into two branches which enter the auditory capsule.

The *Post-auditory nerves* are:

9. The *Glossopharyngei.*

These nerves arise, in common with the next, from the medulla oblongata; and the roots of both leave the skull by an aperture behind the auditory capsule on each side, and form a common ganglion. From this the trunk of the glossopharyngeal is given off. It passes downwards and forwards to the root of the tongue, which it enters and then supplies that organ. Moreover, it gives off muscular branches and a large anastomotic branch to the seventh.

10. The *Pneumogastrici* or *Vagi.*

Immediately after leaving the ganglia these nerves separate from the glossopharyngeal and each gives off a cutaneous branch to the dorsal integument of the head and trunk: it then divides into two branches, one of which (*a.*) runs on the inner side of and above the cutaneous branch of the pulmo-cutaneous artery, the other (*b.*) lies below and diverges from the first.

 a. is the *laryngeal* nerve. It passes beneath the first cervical nerve, then crosses over the third aortic arch and, about its middle, turns sharply round it to be distributed to the larynx. This nerve corresponds with the *recurrent laryngeal* of the higher animals.

 b. is the *splanchnic* branch. It gives off (*gastric*) branches to the gullet and stomach, and a fine nerve (*cardiac*) which passes beneath the pulmonary artery and along the root of the lung to the heart, and ends in ganglia situated in the septum of the auricles. The splanchnic branch finally enlarges and is distributed to the lungs and stomach.

The *myelon* or spinal cord is continued back from the hind-brain as a subcylindrical cord, which lessens somewhat rapidly towards its apparent end at the level of the seventh vertebra. It does not really end here, however, but is continued back as a slender filament, the *filum terminale*, to the commencement of the canal of the urostyle. The diameter of the cord is somewhat enlarged opposite the origin of the nerves for the limbs. In transverse sections, the cord is seen to be not truly cylindrical and to be indented by two longitudinal grooves, one dorsal and one ventral, which leave but a small connecting bridge between its two halves. In the centre of this is a canal, the *canalis centralis*, the cavity of which is continued forwards into the fourth ventricle.

Ten symmetrically disposed pairs of nerves come off from the sides of the cord, each nerve having two roots, one from the dorsal surface of the lateral half of the cord and one from the ventral half. The dorsal root presents a small ganglionic enlargement, beyond which it joins the ventral root to form the common trunk of the spinal nerve. The roots of the hinder spinal nerves are very long and lie, side by side, for some distance, in the spinal canal.

The first spinal nerve leaves the neural canal by the interspace between the arches of the first and second vertebræ, so that there is no suboccipital nerve in the Frog. It gives a branch to the muscles which move the head upon the atlas, but the main trunk of it descends behind the mandible, along with the glossopharyngeal nerve, and is distributed to the muscles of the tongue. It therefore answers to the *hypoglossal nerve* in the higher Vertebrata.

The second and third spinal nerves, of which the second is the larger, unite to form a '*brachial plexus*,' and are distributed chiefly to the fore-limb.

The fourth, fifth and sixth spinal nerves go to the middle parietes of the body.

The seventh, eighth and ninth, are large nerves which unite to form the *lumbosacral plexus*, whence nerves are given off to the posterior parietes of the body, and to the hind-limb. The nerves of the latter are the *crural* to the front part of the thigh, and the *sciatic,* which passes to the back of the thigh and ultimately divides into the *peronæal* and *tibial* nerves which supply the leg and foot.

The tenth spinal nerve leaves the neural canal by the coccygeal foramen, and is distributed to the adjacent parts.

Sympathetic.

The *sympathetic* system consists of ten ganglia, connected by longitudinal commissures, and situated on each side of the ventral face of the vertebral column; in the region of the dorsal aorta they come into close relation with it. Each sympathetic ganglion is joined by a communicating filament with one of the spinal nerves, and the most anterior ganglia are united, in the same way, with the ganglion of the ninth and tenth cerebral nerves. From this a delicate cord, which must be regarded as the most anterior part of the sympathetic, passes into the cranial cavity, on the inner side of the periotic capsule, and unites with the Gasserian ganglion.

The branches of the sympathetic accompany the vessels, and large branches are given to the viscera of the abdomen.

The *Olfactory organs* are two wide sacs which occupy all the space between the mesethmoid cartilage, the antorbital processes, and the premaxillæ and maxillæ, and open in front and dorsally by the external nares, behind and ventrally by the posterior nares. The inner faces of these sacs are lined by a very peculiar epithelium, and the olfactory nerves, with some branches of the trigeminal, are distributed to them.

M.   13

The *Eyeball* is lodged in the orbit and protected by the eyelids described above. It has four *recti* muscles which proceed from the inner wall of the orbit, and are attached to the circumference of the globe; within these is a *retractor* muscle with similar attachments, ensheathing the optic nerve, while two *oblique muscles* proceed from the anterior and inner wall of the orbit and are attached to the dorsal and ventral faces of the bulb. In addition, a fine tendon passes from the outer end of the lower eyelid, or nictitating membrane, and is attached to the fibres of the *retractor bulbi*—the effect of which is that when the bulb is retracted the nictitating membrane is raised over the eye. The upper lid has no muscles. A secretory organ, termed the *Harderian gland*, is situated in the anterior part of the orbit beneath the superior oblique muscle.

The sclerotic is cartilaginous but contains no ossifications, and the lens is nearly spherical. There is no *pecten*.

The *Ear* consists of an essential part—the *membranous labyrinth*—lodged in the periotic capsule, and accessory parts, the *columella auris*, the *tympanic membrane* and the *tympanum*.

The former consists of the three ordinary semicircular canals, with their vestibular dilatations, which open into a vestibule divided into *utriculus* and *sacculus*. The latter, especially, contains a great quantity of white crystalline calcareous otoliths.

On the outer side of the vestibule is a small dilatation which is possibly a rudimentary *cochlea*.

The membranous labyrinth is contained in the partly cartilaginous, partly osseous, periotic capsule into which it fits but loosely; the interval is filled with a fluid, the *perilymph*. In the outer face of the periotic capsule is an oval opening, the *fenestra ovalis*, into which the end of the *columella auris* fits. This columella is shaped like a pestle, the end of the handle of which is fitted with a cross-piece. The rounded inner end of the pestle, which is fixed by fibrous tissue into the fenestra ovalis, is cartilaginous. The middle of the handle is ensheathed in bone, while the outer part is cartilaginous. The cross-piece is fixed into the inner face of the membrana tympani, which is lined externally by the integument, internally by mucous membrane, continuous with that of the mouth through the Eustachian recess. The mucous membrane of the tympanic cavity covers only the ventral face of the columella, over the dorsal face of which the posterior division of the facial nerve passes.

The *Tongue.* This organ, as has been seen, is fixed only in front to the mandible, and by the anterior half of its ventral aspect to the floor of the mouth; the posterior half being free and bifid at the extremity. Narrow-ended and broad-ended papillæ (*papillæ filiformes* and *fungiformes*) are scattered over the whole dorsal aspect of the tongue and are largest in front; small glands lie between these papillæ.

The fungiform papillæ contain the ultimate ramifications of the glossopharyngeal nerve, and the epithelium covering their summits is peculiarly modified.

The *Integument.* No special organs of touch have been observed, but the integument is remarkable for the immense number of close-set simple glandular cæca which open upon its surface. In the swollen integument which covers the base of the inner digit in the males, large papillæ with interposed glands are developed.

A singular body of unknown function, the *browspot* or *inter-ocular* gland, consisting of a spheroidal sac with minute

13—2

cells, occurs in the integument of the frontal region of the head.

Cells containing pigment abound in the integument and undergo remarkable changes of form, the pigment being sometimes drawn together into a spheroidal mass—at other times distributed in a radiating fashion.

LABORATORY WORK.

A. GENERAL STRUCTURE.

1. Go over the specific characters given above (p. 164).

2. The divisions of the body : *head, trunk,* two pairs of *limbs* (see p. 159).

   a. The head.

Somewhat triangular, with the blunted apex turned forwards and passing broadly, without any neck-constriction, into the trunk ; notice the prominent eyes with their lids ; the *membrana tympani,* a part of the integument stretched over a hard ring, placed on each side, behind and somewhat below the eyes ; the two apertures of the nostrils (*anterior nares*) between the eyes and the end of the snout ; the *mouth opening;* the hard parts felt through the skin on the upper side of the head ; the soft flexible throat.

Pass a bristle into one of the anterior nares. Make a small opening in one of the tympanic membranes and pass another bristle into it. Now open the mouth widely ; and, if the bristles have been thrust far enough, the end of the former will be seen traversing the posterior nasal opening in the roof of the mouth : while the end of the other will appear in the Eustachian recess which lies at the sides of the back of the oral cavity. The fleshy tongue will be seen, with its bifurcated free end turned backwards. Turn it forwards to see the attachment of its base to the floor of the mouth and to the front part of the lower jaw. Notice the slit of the *glottis* in the hinder part of the floor of the mouth, and above this the opening of the œsophagus. Pass a bristle into the former, and a probe into the latter. Notice the fine teeth in the upper jaw and on the palate.

b. The trunk.

Tapering towards the hinder end; and allowing the hard parts of the skeleton to be felt beneath the soft integument on the dorsal side, and in the anterior half of the ventral aspect; rounded and soft on the greater part of the sides and belly; the *cloacal aperture* near the dorsal surface of the posterior end of the trunk.

c. The limbs.

α.   The anterior pair; their three subdivisions, *brachium, antebrachium,* and *manus;* the four *digits.*

β.   The posterior pair; their length as compared with that of the anterior; their subdivision into *femur, crus,* and *pes:* the five long *digits;* the well-developed *web;* the horny prominence (see p. 161).

3.   Raise the integument of the abdomen with forceps and slit it open with scissors from the lower jaw to the origin of the hind limbs, a little on one side of the middle line. Observe the spacious lymph cavities

between the skin and the subjacent muscular wall of the abdomen; also a vein which occupies the middle line of the inner face of this wall and is usually visible through it.

4. Raise the muscular wall of the abdomen and cut it in the same way, a little on one side of the middle line, sufficiently to lay open the abdominal cavity, taking great care to avoid the bladder which lies at the posterior end of the cavity. Note the conspicuous vein (*anterior abdominal*) which lies beneath the muscles in the middle line of the belly. The liver, stomach and intestines will be seen; at the sides, in the female, the ovaries and oviducts will be very conspicuous in the breeding season. Insert a small blow-pipe into the cloacal opening: air blown in will distend the large bilobed urinary bladder. If the lungs are distended with air, one will be visible on each side of the anterior end of the abdominal cavity, and the extremity of the bristle passed into one of them, through the glottis, will be seen. Lay open the stomach to see the end of the probe passed into the œsophagus.

By turning the intestines on one side, the kidney, the *corpus adiposum* and the testis (in the male) will be exposed. Notice a number of small white patches on each side of the vertebral column. They are accumulations of calcareous crystals.

5. In front of the liver, the apex of the heart will be seen through the pericardium. Lay the latter open and observe the position of the heart.

6. Cut away the left fore-limb and the left hind-limb, with so much of the left half of the vertebral column

and skull as is needful to lay open the cavity which contains the cerebro-spinal nervous centres. Pin the frog in a dissecting dish, on its right side, with sufficient water to cover it, and study the position of the various organs in relation to a median longitudinal plane, making a careful diagram of the parts displayed.

7. In a frog which has lain in bone-softening solution (say 1% chromic acid) sufficiently long to soften the bones, make transverse sections (1) through the eyes, (2) through the centres of the tympanic membranes, (3) through the shoulder-girdle, (4) through the hinder half of the abdomen. Compare them with the foregoing dissection.

B. DISSECTION OF THE VISCERA IN THE VENTRAL CAVITY.

1. Lay a frog, which has been killed with chloroform, on its back and pin it out on a layer of paraffin or beeswax, under water; divide the skin along the abdominal median line from the pelvis to the front of the lower jaw; next make a transverse incision at each end of the longitudinal one, and then throw outwards the two flaps of skin thus marked out. The following points may now be noted.

   a. A great vein (*musculo-cutaneous*) on the undersurface of each flap of skin, about the level of the shoulder.

   b. Some of the muscles of the abdominal wall, covered by a thin aponeurosis: through this latter can be seen—

      a. The *rectus abdominis* running from pelvis to

sternum close to the middle line, and divided into a number of bellies by transverse tendinous intersections.

β.   Other muscles outside the rectus on each side.

c.   The *pectoral region;* part of its hard parts in the middle line, only covered by tendinous tissue: external to this, muscles running towards the shoulder-joint.

d.   The *muscles of the throat;* small and with a general direction from the lower jaw towards the sternum and shoulder-girdle.

2.   Raise the tissues of the body-wall with a pair of forceps and carefully divide them, a little to the right of the median line, so as to open the body-cavity without injuring its contents; prolong the incision from the pelvis to the posterior end of the breast-bone; make a transverse incision close to the pelvis and throw back the flap on each side: on the deep side of the left flap will be seen a large vein (*anterior abdominal*).

Seize the posterior border of the sternum with a pair of forceps and raise it up: on looking beneath it several fibrous bands will be seen running from it and the hard parts in front of it to subjacent parts; carefully divide these: then, with a strong pair of scissors, cut through the hard parts in the median line, being very careful not to injure the organs beneath them; turn each half outwards and pin it in that position; stretch out each fore-limb to its fullest extent and fasten it with pins.

3. Note the smooth moist membrane (*pleuroperitoneum*) lining the inside of the body-cavity and covering the outside of the contained viscera.

4. The *liver* will be readily recognized as a great brownish mass covering a great part of the other abdominal viscera: lying in a cleft in its anterior border and partly concealed by it, will be seen a delicate sac in which pulsations are going on; this sac is the *pericardium:* if it be opened and removed very carefully, the heart and some of the great blood-vessels will be laid bare; clean carefully the two great trunks (*aortic arches*) which diverge from the anterior end of the heart, following each to the point of its division into three vessels.

5. **The heart.**

   a. Note the general form of the organ.

      a. Its posterior conical thick-walled portion (*ventricle*) with the apex turned backwards.

      b. The *truncus arteriosus:* a sub-cylindrical part, arising from the right side of the base of the ventricle and dividing anteriorly into the two aortic arches.

      c. The *atrium:* thin-walled, rounded, lies on the dorsal aspect of the truncus and ventricle. The separation between the two auricles is not visible externally.

      d. Carefully raise the ventricle: lying beneath it (that is, on its dorsal side) will be seen another division of the heart, the *sinus venosus;* it lies
      . between the atrium and the great systemic

veins, which enter it through the dorsal and posterior walls of the pericardium.

c. The further examination of the structure of the Frog's heart requires a good deal of care and the use of a lens of low magnifying power. In a chloroformed Frog the heart is distended with blood when it ceases to beat. When all signs of contractility have disappeared, the distended heart should be removed from the body by cutting through the adjacent parts in such a manner as to leave the terminations of the veins and the origins of the aortic trunks intact. The organ should next be transferred to a shallow dissecting dish and covered with weak spirit. The right and left walls of the atrium being now carefully slit and the blood which they contain washed away, the delicate septum of the auricles will become visible. By cautiously removing the ventral face of the ventricle, its cavity will be laid open and the auriculo-ventricular opening will be displayed. If the ventral wall of the truncus arteriosus is laid open longitudinally with fine scissors, the valves in its interior will become visible. The pulmonary vein runs along the dorsal aspect of the sinus venosus, between the right and left superior venæ cavæ, to the left auricle, into which it opens close to the dorsal attachment of the septum.

The natural relations of the different divisions of the heart should be carefully studied in such a dissection as is described in A. 6.

b. *The pulsation of the heart.* This should be studied in a Frog rendered insensible by chloroform
or by being pithed; though the latter operation
causes such dilatation of the vessels that little or
no blood may afterwards flow through the heart,
yet the organ goes on beating.

   a. Watch the movement carefully; it is a regularly alternating series of contractions and dilatations.

   b. It will be seen that the two auricles contract
together; immediately after them, the ventricle;
and then, instantly, the bulbus arteriosus.

   c. Raise the ventricle so as to see the venous
sinus; note that it contracts immediately before
the auricles.

6. **The parts exposed by the preceding dissections**
(B. 1. 2).

Draw them carefully without disturbing them.

   a. The *throat-muscles:* through the broad thin muscle in front (*mylo-hyoid*) is seen the *hypoglossal
nerve.*

   b. The *larynx:* forming a hard prominence in the
middle line, just in front of the aortic arches.

   c. The *heart* and *aortic arches* (see B. 5. i.): the
three terminal branches of the latter, viz.—

      a. The *carotid trunk;* the anterior division; end-
in a small reddish body (the *carotid gland*).

      β. The *systemic aortic arch.*

      γ. The *pulmo-cutaneous artery:* the hindmost
branch.

d. The *liver:* a great brown two-lobed mass; its left lobe the larger and subdivided into two.

e. The *lungs:* the posterior ends of these may be seen as sacculated pouches, one on each side of the liver, but they are frequently not visible until the latter organ has been removed.

f. The *stomach:* a small portion of this is seen projecting beyond the lower left border of the liver.

g. The *intestine:* a convoluted tube, continuous with the stomach, and slung by a delicate membrane, the *mesentery:* posteriorly the intestine ends in a dilated portion (*rectum*) which runs into the pelvis.

h. The *urinary bladder:* a thin-walled bilobed sac (which may or may not be distended) appearing just in front of the pelvis.

i. The *fat masses:* long slender yellow processes appearing on each side of the liver.

In *R. temporaria*, the urinary bladder is much more deeply lobed and also much larger proportionately, than in *R. esculenta.*

7. **The liver.**

a. Study its form more closely. (6. *d.*)

b. Raise its lower border; between its two lobes will be seen a small greenish sac, the *gall-bladder*.

c. Carefully cut away the liver, except its deepest part, close to the venous sinus.

d. Tease out a bit of liver in 0·75% sodic chloride solution and examine with ⅛ obj.

α. Numbers of polygonal granular cells (*hepatic cells*), with oil-drops in them, will be seen.

β. Treat with acetic acid: a nucleus, or sometimes two, will be rendered apparent in each of the cells,

8. **The stomach, intestine, pancreas and spleen.**

    a. Cut away the front of the pelvis with a stout pair of scissors, taking care not to injure the urinary bladder: pass a probe from the anus, through the *cloaca*, into the rectum: uncoil the intestine and spread out the mesentery, so far as is possible without cutting the latter.

    a. The *spleen:* a small red body lying in the mesentery, near its attachment to the back of the abdomen.

    β. The *stomach:* an elongated sac on the left side of the abdominal cavity: the narrower tube (*œsophagus*) opening into its anterior end.

    γ. The *intestine:* its length and varying diameter; especially the great width of its rectal portion: its posterior termination in the cloaca.

    δ. The *pancreas:* a pale-coloured compact mass lying in the mesentery near the commencement of the intestine.

    ε. The *bile-duct.* Slit open the duodenum where the right end of the pancreas is attached to it: a small aperture will be seen on the mucous membrane of the intestine at this point; this is the opening of the bile-duct: pass a bristle into it.

    ζ. The *mesentery:* its width; mode of attachment to the intestine; the blood-vessels run-ing in it.

b. Divide the œsophagus close to the stomach and the rectum near the cloaca: remove all the por-tion of alimentary canal between these two points, cutting through the mesentery.

α. Pass a probe up the œsophagus into the mouth.

β. Open the upper end of the intestine and snip off a bit of its internal layer (mucous mem-brane) and mount in normal saline solution: examine with $\frac{1}{8}$ obj.; on the fragment will be found minute prominences (representing the *villi* of the higher animals) covered by a closely-set layer of cells (*epithelium*).

9. **The kidneys.** These organs are now exposed as two elongated deep red bodies lying in the posterior part of the perivisceral space close to the verte-bral column; clear away any bits of mesentery, &c. which may cover them; note—

a. The duct—*ureter* (female) or *genito-urinary canal* (male)—running from the outer side of the pos-terior part of each kidney to the cloaca. Open the cloaca and pass a bristle into the opening of one of the ureters.

b. In the male *R. esculenta* each duct is somewhat dilated after leaving the kidney: it then narrows again and opens on the posterior surface of the cloaca by an oblique slit with sharply defined edges. In *R. temporaria* the duct does not dilate, or only very slightly; but on its outer side lies a glandular mass (*vesicula seminalis*), from the inner side of which a number of minute ducts open into the genito-urinary canal. The aperture of the latter in the cloaca is round and has tumid edges. In the female of both species the ureters are very slender.

c.   The vein (*renal portal*) entering the kidney at its posterior outer border.

d.   The great vein (*vena cava inferior*) lying between the kidneys and chiefly formed by their efferent (*renal*) veins.

Now that the cloaca is open, trace the opening of the urinary bladder (B. 6. *h*) into it.

10.   **The generative organs.**

    a.   In the male.

        α.   *The testes:* a pair of yellowish bodies lying in front of the anterior ends of the kidneys; their form.

        β.   The ducts of each testicle (*vasa efferentia*) entering the inner border of the kidney of the same side in order to communicate with the genito-urinary canal (9. *a*).

        γ.   Remove the testes : open one and press out some of its contents upon a slide and mount in common water: examine with $\frac{1}{6}$ objective.

The *spermatozoa:* bodies provided with a small oval *head* and a long vibratile *tail* in *R. esculenta :* in *R. temporaria* the oval head is absent : their movements.

    b.   In the female.

        α.   *The ovary :* an organ varying much in size

with the season of the year. The numerous *ova* in it.

β.   The *oviduct:* a convoluted tube, not continuous with the ovary, and running back to open into the cloaca. The greater part of the oviduct is opaque and glandular; the part near the cloaca however is dilated, thin-walled and transparent. The oviducts open on the posterior wall of the cloaca a little anterior to the openings of the ureters.

11.   **The mouth, œsophagus, and respiratory organs.**

a.   Open the mouth: note on its roof near the front the two hinder openings of the nasal cavities (*posterior nares*); farther back, the two larger openings of the *Eustachian recesses;* the long *tongue* in the floor of the mouth; fixed by its front end to the lower jaw; its free bifid end turned towards the throat.

b.   Enlarge the mouth-opening by cutting through the sides of the buccal cavity with a pair of scissors: pull down the lower jaw so as to see the chamber (*pharynx*) behind the mouth.

c.   On the floor of the pharynx is a narrow opening (*the glottis*): pass a probe down it through the *larynx* and very short *trachea* into the lungs.

d.   Remove the lungs; open one; it is a thin-walled cavity with a sacculated inner surface.

e.   Trace the œsophagus up to the pharynx.

C.   THE CIRCULATION OF THE BLOOD IN THE FROG'S WEB.

1.   Get a piece of thin board, about 5 inches long and 2¼ broad; in the middle of one end of it cut

a V-shaped notch about the size of a spread-out frog's web : place the frog on the board, belly downwards, and fix it by passing round it two or three turns of tape : next tie threads round the toes of one hind-foot, and by means of them spread out the web over the notch in the board, taking great care that it is only very lightly stretched. The animal should be kept moist by a bit of wet blotting-paper spread over its back.

2. Examine the web with 1 inch obj. : Note—

a. The black *pigment-cells* in the skin; sometimes irregularly branched ; sometimes more compact.

b. The close *network of blood-vessels* lying deeper than the pigment-cell layer.

  a. The *arteries*, running mainly towards the free edge of the web, and constantly diminishing in size as they give off branches ; the blood-flow in them from larger to smaller branches.

  β. The *capillaries*, in which the arterial branches end : small vessels forming a close network and frequently branching or anastomosing without much altering their size.

  γ. The *veins*, formed by the ultimate union of the capillaries, and increasing in size by union with one another ; the blood-flow in them from smaller to larger trunks.

c. The *blood-flow :* the current being marked by the solid bodies (*corpuscles*) carried along in the fluid : it is most rapid in the arteries ; slowest, and most uniform, in the capillaries.

M.   14

3. Place a small drop of water on a bit of a thin cover-glass, and place the bit, with the water downwards, gently on the web: then examine the following points with $\frac{1}{4}$ or $\frac{1}{8}$ obj.; note—

a. The *walls of arteries, capillaries, and veins.*

   α. The arterial walls, tolerably thick, seen as a clear well-defined band on each side of the blood-stream.

   β. The capillary walls; difficult to see; merely a thin somewhat more transparent boundary line.

   γ. The venous walls; much like the arterial.

b. *The blood-flow in the small arteries of the web.*

   α. The rapid stream in the middle, containing most of the red corpuscles.

   β. The slower stream along the edge (*inert layer*), containing many colourless corpuscles.

c. *The flow in the capillaries:* much slower than in the arteries; the frequent distortion of the red corpuscles in the capillaries from pressure, &c.; their elasticity as indicated by the readiness with which they recover their shape when the cause of distortion is removed; the way the white corpuscles creep along, with a tendency to stick to the capillary wall.

4. Examine a drop of frog's blood with the microscope ($\frac{1}{4}$ or $\frac{1}{8}$ obj.). Sufficient blood to supply a whole class for this purpose can be obtained by killing one frog and opening its heart.

It consists of solid bodies (*corpuscles*) floating in fluid (*plasma*).

a. The red corpuscles.

α. *Their form :* oval when seen in front face; almost linear in profile but slightly swollen at the centre.

β. *Their size :* their length, breadth, and thickness; measure.

γ. *Their colour :* pale yellow, when seen individually; redder if a thick mass of them is looked at.

δ. *Their structure :* they are homogeneous for the most part, but possess a round granular central nucleus.

ε. Treat with water; they swell up and become more spherical; their colouring matter is gradually discharged; the nucleus is rendered very evident, and ultimately all the rest of the corpuscle disappears.

ζ. Treat with dilute acetic acid; results same as with water, but produced more rapidly.

b.   The white corpuscles.

Less numerous than the red : their colour, size, granular character, nucleus, and changes of form (*amœboid movements*): see III. B.

D. THE EXAMINATION OF A PREPARED SKELETON.

The skeleton of a Frog may be prepared for examination by removing the viscera from the body, and roughly dissecting away the muscles, &c. Then place the remainder in water and let it macerate for about a week; afterwards carefully pick away the soft parts, with forceps, from the bones and cartilages.

14—2

a. Its general arrangement.

1. The central axis, consisting of the *vertebral* or *spinal column* and of the *central parts of the skull* which lie, in front of the spinal column, in the same antero-posterior line.

2. Lateral parts, supported, directly or indirectly, by the axis.

    a. The appendages proper. (*Limbs* and *limb-arches*.)

    a. The *fore-limbs :* their supporting *shoulder-girdle* or *pectoral arch*, not directly attached to the axial column ; the *limb proper ;* its main divisions ; *humerus, radius* and *ulna* (the two latter ankylosed), *carpus* and *digits*.

    β. The *hind-limbs :* their supporting arch (*pelvic girdle*), carried directly by bony processes proceeding from the vertebral column ; the *limb proper ;* its main divisions, *os femoris, tibia* and *fibula* (ankylosed), *tarsus, digits.*

    b. The facial and lateral cranial bones.

b. The vertebral column.

It consists of an anterior segmented portion (each segment being a *vertebra*) and of a posterior unsegmented portion (*the urostyle*).

1. Examine carefully and draw various aspects of a detached *vertebra*, say the third.

    a. Its solid flattened ventral part *centrum*) ; with an anterior concave and a posterior convex surface.

    b. The *neural arch :* an arch of bone springing from the sides of the dorsal aspect of the centrum of the vertebra and not quite so wide as the centrum in its antero-posterior diameter.

α. The *transverse process:* a bony bar on each side, arising from the arch and passing outwards and a little downwards.

β. The *articular processes (zygapophyses)*; an anterior and posterior pair, springing from the sides of the arch ; the *anterior*, having their smooth articular surfaces directed upwards ; the *posterior* with similar surfaces directed downwards. ,

γ. The short *spinous process*, springing from the dorsal aspect of the arch and directed backwards.

c. The *neural canal* enclosed between the arch and the body.

2. Examine the remaining vertebræ.

a. The first vertebra (*atlas*): its *body* produced forwards into a wedge-shaped process which lies between the occipital condyles : its *arch*, sometimes incompletely ossified in the region of the spinous process, which is rudimentary; posterior zygapophyses only present : the large concave anterior facets, partly on the arch and partly on the centrum, with which the skull articulates.

b. The 2nd, 4th, 5th, 6th and 7th vertebræ : closely resembling the third, the chief differences being found in the varying size and direction of the transverse processes, all of which are smaller than those of the third vertebra.

    c.   The 8th vertebra: the concave facet at each.end of its *centrum.*

    d.   The 9th vertebra (*sacrum*): its *centrum;* convex in front and with two convex tubercles behind : its large strong *transverse processes* (*sacral ribs*) directed somewhat backwards and expanded at their ends.

3.   The posterior unsegmented portion of the vertebral column (*urostyle*).

    a.   An elongated rod-like bone, rather thicker towards its anterior end, which bears two concavities.

    b.   Its posterior end ; tubular in the dried skeleton, but in the fresh state filled with a cartilage which projects beyond it posteriorly.

    c.   The prominent ridge along its dorsal surface ; wider and higher in front ; getting thinner and gradually disappearing towards the posterior end.

    d.   The small canal contained in the forepart of the ridge.

    e.   The two minute passages leading from the canal to the outside of the ridge on each side.

4.   The vertebral column as a whole.

    a.   *Its composition.*

        α.   Its anterior segmented part formed of the nine vertebræ.

        β.   Its posterior unsegmented part formed by the urostyle and nearly as long as the segmented part.

b. *Its ventral surface.*

    α.   The solid bony axis formed by the bodies of the vertebræ in front and continued posteriorly by the ventral rounded part of the urostyle.

    β.   The *transverse processes :* their size and direction : those of the second vertebra, slender and directed nearly straight outwards; those of the third and fourth, largest of all and with a distinct inclination backwards : those of the fifth, sixth, seventh and eighth vertebræ smaller than the rest and directed nearly outwards ; those of the ninth, very stout, directed outwards upwards and backwards, and having the iliac bones attached to their distal ends. The transverse processes arise from their arches closer to the centrum in the second, third and fourth vertebræ than in the others.

    γ.   The *inter-vertebral foramina :* the interspaces left between each pair of neural arches, below the level of the zygapophyses.

c. *The dorsal surface of the spinal column.*

    α.   The ridge down its middle formed by the row of spinous processes, and continued behind by the dorsal ridge of the urostyle.

    β.   The lateral prominences caused by the articular processes : the articulations between the anterior articular processes of one vertebra, and the corresponding posterior processes of its predecessor.

    γ.   The spaces left between the dorsal part of

each neural arch and its successor : between the atlas and the second vertebra and between the eighth and ninth vertebræ; these are almost obliterated by the approach of the respective arches.

d. The *neural canal.*

α. Closed by the centra of the vertebræ beneath, and incompletely by the neural arches on the sides and above : its backward continuation as the canal in the front part of the ridge of the urostyle.

β. The communication with it, of the intervertebral foramina (4. *b.* γ), and the dorsal intervals between the neural arches (4. *c.* γ) : also of the openings in the urostyle (3. *e*).

c. **The skull.** The prepared bony skull of the frog is difficult to understand, for two reasons; firstly, on account of the dried-up condition of the cartilages of which, in the fresh state, it is in part composed; and, secondly, on account of the tendency of many of its constituent bones to become ankylosed together in the adult : the following points can however be made out with tolerable ease. Drawings should be made of each aspect of the skull.

1. Examine the *posterior end of the skull.*

a. The large aperture (*foramen magnum*) in the middle line, leading into the *cranial cavity.*

b. The convex surface (*occipital condyle*), on each side of the foramen magnum, which articulates with the corresponding concave facet on the front of the atlas.

c. The bone bearing the condyle on each side and, with its fellow, enclosing the foramen magnum, is the *exoccipital.*

d. The thick bone running outwards in front of the exoccipital, on each side, protects the front part of the internal ear, and is the *pro-otic* bone.

e. Between these two bones, on the outer side of the chamber which contains the organ of hearing (*periotic capsule*), is a cartilaginous interspace containing an oval aperture, the *fenestra ovalis.* In this is fixed the inner end of a partly cartilaginous and partly osseous rod, the *columella auris.*

f. Attached to the outer end of the pro-otic bone is a hammer-shaped bone—the *squamosal,* which extends from the pro-otic bone to the articulation of the lower jaw.

2. *The roof of the skull.*

a. Passing forwards from the exoccipitals are two long flat bones, the *parieto-frontals,* one on each side of a median suture which answers to the sagittal and frontal sutures in man.

b. In front of these come two triangular bones— *the nasals.*

c. In front of the nasals are two other bones, which belong rather to the ventral than to the dorsal face of the skull. They form the extreme front of the snout, and each sends a process towards the nasals; these are the *premaxillary* bones.

3. *The base of the skull.*

a. Running along the greater part of the floor of the cranial cavity, from the occipital foramen to the vomers, is a bone shaped like a dagger with a short handle and a strong guard. The latter spreads out under the pro-otics. This is the *parasphenoid bone.*

b. Appearing at the base of the skull, at the front end of t'' c parasphenoid, is the *girdle-bone* or *sphenethmoid* (which represents several bones joined together); this bone closes in the floor and sides of the forepart of the cranial cavity and also its roof, being concealed in the latter place by the anterior ends of the parieto-frontals. The sphenethmoid has a single cavity behind, which enters into the formation of the cephalic chamber, and two cavities in front, one for each nasal chamber, separated by a septum.

c. Running out transversely from the girdle-bone and the anterior end of the blade of the parasphenoid on each side, is the slender *palatine.*

d. In front of the end of the blade of the parasphenoid and of the palatines are two broad irregularly shaped bones, each bearing an oblique row of teeth on its posterior part: these are the *vomers.*

e. The middle anterior boundary of the contour of the skull, in this view, is formed by the dentigerous parts of the pre-maxillæ; and, behind them, by the maxillæ and the quadrato-jugal bones (4. *a. b*). Running backwards from the outer end of the palatine, and closely applied to the maxilla, is a bone, which soon separates from the maxilla, and becoming broad and stout, bifurcates; the inner process nearly joins the parasphenoid and is moveably articulated with the skull; the outer runs along the inner face of a cartilage (the *suspensorium*), on the outer face of which the squamosal rests: this is the *pterygoid bone.*

At the outer edge of each vomer, immediately in front of the palatine, is an aperture which leads into the *nasal cavity.* These two apertures are the *posterior nares.*

4. *The side of the skull.*

a.   Running back from that part of the pre-maxilla which bounds the gape is a long bone, forming almost the whole of the rest of the upper edge of the gape; this is the *maxilla.*

b.   Joined to the posterior end of the maxilla is a small bone, which at its posterior end is attached to the distal portion of the squamosal. This is the *quadrato-jugal.*

c.   The under jaw or *mandible* consists of two distinct portions, or *rami*, which meet in the middle line in front, and which, behind, articulate with the extremities of the suspensorial cartilages. In the articular end of each suspensorial cartilage there is an ossification which represents the *quadrate* bone in other Vertebrata, and is united with the jugal to form the quadrato-jugal.

In each *ramus* three pieces may be made out—

α. A central axis formed of cartilage (*Meckel's cartilage*), which enlarges at its posterior end in order to articulate with the suspensorial cartilage, while at the opposite or symphysial end it is ossified to form the *mento-Meckelian* bone.

β. A posterior inferior piece, which runs nearly to the middle line in front (*angulo-splenial*) and partly ensheaths the foregoing.

γ. A small anterior superior piece (*dentary*).

d. **The hyoid bone or cartilage.**

α. Its broad somewhat tetragonal central part (*body*), bearing a number of processes, viz.—

β. The *anterior cornua*, proceeding from the front of the body on each side : each is a long slender curved cartilage running at first forwards, then backwards and outwards, and finally forwards and upwards, to become attached to the periotic capsule beneath the fenestra ovalis.

γ. The *posterior cornua*, or *thyro-hyals;* bony, and shorter and thicker than the anterior cornua: attached to the posterior border of the body near the middle line and diverging as they run backwards.

δ. Two pairs of smaller processes formed by the elongation of the anterior and posterior angles of the body of the hyoid.

e. **The sternum and shoulder-girdle.**

1. *Their general arrangement :* they form an incomplete ring round the fore-part of the trunk : this ring is

composed partly of bone, partly of cartilage. Note the hollow (*glenoid fossa*) with which the fore-limb articulates.

a.   *The sternum :* situate in the ventral median line and made up of several parts: beginning behind we find—

    α.   The *xiphisternum;* a thin cartilage, wide behind, narrow in front, where it passes into—

    β.   A median cartilage ensheathed in bone, the *sternum* proper. The anterior end of the sternum unites with the posterior and internal angles of the coracoids (*b.* γ), the inner edges of which meeting in the middle line separate the sternum from—

    γ.   The *omosternum*, consisting of a slender, flattened bone, terminated in front by an expanded cartilage. The posterior end articulates with the præcoracoids and the clavicles.

b.   The *shoulder-girdle :* beginning dorsally, it exhibits on each side—

    α.   A thin expanded dorsal portion, partly cartilaginous, partly ossified: the *supra-scapula.*

    β.   Next a bony segment—the *scapula*, the posterior inferior edge of which is excavated by the *glenoidal fossa.*

The ventral parts of the shoulder-girdle, which lie between the two scapulæ, meet in the middle line; the part on each side is subdivided into an *anterior* and a *posterior portion* by a large foramen.

γ. The bony piece running behind the foramen, from the scapula almost to the middle line, is the *coracoid*. Where the coracoid unites with the scapula it contributes to the formation of the glenoidal cavity.

δ. The adjacent margins of the two coracoids are fringed with cartilage (*epicoracoid*), which passes in front of the foramen into a bar of cartilage—the *præcoracoid*—which, externally, is continuous with the cartilage which lies tween the scapula and coracoid and helps to bound the glenoidal cavity.

ε. Closely attached to the fore-part of the præcoracoid lies a bone, the *clavicle*, the outer end of which articulates with the coracoid and scapula, the inner with the omosternum.

. Draw the whole pectoral arch carefully, shading differently the bones and cartilages.

f. The bones of the fore-limb.

a. The *arm-bone* (*humerus*).

α. A somewhat cylindrical bone, with an *articular expansion* at each end and a *shaft* uniting them.

β. The great ridge (*deltoid crest*) on its antero-internal surface, to which a muscle was attached.

The development of this crest is greater in the males than in the females.

b.   The *bone of the forearm.*

  a.   Hollowed out above to fit the lower end of the humerus.

  β.   Shewing below a tendency to divide into the two bones of which it is made up; viz. the *radius* and the *ulna.* When the limb is stretched out at right angles to the body with the pollex forwards, the radius is on the anterior, and the ulna on the posterior side of the axis of the limb.

c.   The *carpus.* Two bones (*a*, *b*) articulate with the ankylosed radius and ulna. A third bone (*c*), on the radial side of the carpus, articulates only with the carpal bones on the proximal and distal sides of it. A large bone (*d*) occupies two-thirds of the ulnar side of the carpus, and articulates with *a*, *b* and *c* on one side, and with the third, fourth and fifth metacarpals on the other. Two small ossicles articulate with the distal face of *c* and bear the first and second metacarpals.

d.   *The digits.*

  Five in number, the first (radial one) being, however, rudimentary: beginning at the ulnar side, we find—

  a.   The fifth digit (that on the outer or ulnar side of the limb): it presents a cylindrical proximal bone (*metacarpal*) followed by three others (*phalanges*), each shorter than its predecessor.

  β.   The fourth digit: a *metacarpal bone* and three *phalanges.*

γ.   The third digit: a *metacarpal bone* with two *phalanges.*

δ.   The second digit: a *metacarpal bone* with two *phalanges.*

ε.   The· first digit (*pollex*) consists only of a small *metacarpal bone.*

g. The pelvic girdle.

a.   Its general form: V-shaped, with the apex turned backwards.

b.   The concavity (*acetabulum*) on each side with which the thigh-bone articulates.

c.   The triradiate fissure running through the acetabulum and dividing each half of the pelvis into three pieces, viz.—

     a.   An anterior elongated piece (*ilium*); subcylindrical in front, where it articulates with the sacrum; bearing behind, on its dorsal aspect, a laterally compressed ridge of bone (*crista ilii*); it forms almost half of the acetabulum.

     β.   A posterior, irregularly rounded piece (*ischium*); closely united posteriorly with its fellow, in the middle line.

     γ.   A small triangular piece (*os pubis*), (*calcified cartilage*—except in old frogs), wedged in between ilium and ischium, and meeting its fellow in the median ventral line, thus forming the *symphysis pubis.*

h. The bones of the lower limb

a.   The *thigh-bone (os femoris)*; its long cylindrical shaft and expanded articular extremities.

b.   The *leg-bone (os cruris).*

α. A very long cylindrical bone, expanded at each end.

β. The grooves on it; one running along the whole ventral surface, but most marked near the ends;. other grooves on the dorsal surface, one at the upper, another at the lower extremity: these indicate that the os cruris is really made up of two united bones, the *fibula* and the *tibia.* When the limb is stretched out at right angles to the body, the tibia is anterior, corresponding with the radius; and the fibula, posterior, correspond-. ing with the ulna.

c. The *tarsus.*

α. Two elongated bones (separate in the middle but united by confluence of their cartilaginous extremities) articulate with the ankylosed tibia and fibula; the anterior, or tibial, of these is the *astragalus;* the posterior, or fibular, the *calcaneum.*

β. With the distal ends of these, two partially ossified cartilages, one on the calcaneal and the other on the astragalar side, articulate. The latter is connected by ligamentous fibres, within which a nodule of cartilage may be found, with the first and second metatarsals, and supports the *calcar* (*d. ζ*).

d. *The digits.* Five in number; the internal one the shortest, the fourth the longest. Their composition—

α. The first, or *hallux* (the most internal); a *metatarsal bone,* followed by two *phalanges.*

β.   The second: same as a, but longer.

γ.   The third: a metatarsal bone with three pha-
      langes.

δ.   The fourth: a metatarsal bone and four
      phalanges.

ε.   The fifth: like the third, but a little shorter.

ζ.   On the anterior or tibial edge of the foot there
      are two, small, more or less cartilaginous,
      ossicles, articulated with the tarsus (*c. β*) in
      such a manner as to resemble an extra digit.
      This *calcar*, or spur, supports the horny pro-
      minence referred to above.

E.   DISSECTION OF THE FROG'S HIND-LIMB TO ILLUSTRATE
      ITS MYOLOGY.

        (For the following dissection it is desirable to have a
frog which has been lying some time in spirit.)

1.   Lay the animal on its back, and make an incision
      through the skin on the front of the limb from the
      symphysis pubis to the ankle ; then reflect the skin
      to each side so as to lay bare the parts beneath it;
      in turning it back note the loose bands and fibres
      (*subcutaneous areolar tissue*) with large *lymph-spaces*
      between them, which unite the skin to subjacent
      parts, and which have to be cut through.

        A number of *muscles* will now be exposed on the
front of the thigh and leg.

2.   **The superficial muscles on the front of the thigh.**
      Separate these gently from one another, tearing
      through the *connective tissue* which unites them.

a.   Each is chiefly made up of a mass, the *belly* of the muscle, which is nearly white and readily tears into bundles in a muscle which has been in spirit; but is softer, redder, and does not so easily split up in a fresh muscle.

b.   At both ends, in most cases, the belly is replaced by dense shiny tissue forming a *tendon.*

c.   The tendons are fixed directly or indirectly to some of the neighbouring bones, the less moveable attachment being the *origin* of the muscles; the point of attachment to the more moveable bone, its *insertion.*

d.   The names of the muscles laid bare on the front of the thigh, are—

   a.   The *sartorius:* a thin flat riband-like muscle running down the middle; it *arises* from the symphysis pubis and is *inserted* into a tendinous expansion (*aponeurosis*) on the inner side of the knee-joint.

   β.   The *adductor magnus:* it becomes superficial along the upper two-thirds of the inner border of the sartorius.

   γ.   The *adductor brevis:* a little bit of it is seen on the inner side of the adductor magnus, close to the symphysis pubis.

   δ.   The *rectus internus major:* a large muscle running along the whole inner side of the thigh; arises from the symphysis pubis below the sartorius and is inserted into the same aponeurosis as that muscle.

   ε.   The *rectus internus minor:* a thin muscle lying

inside and rather behind the rectus internus major. It arises from the pelvis close to the anus and is inserted into the aponeurosis about the knee-joint.

ζ.   The *adductor longus:* it is partly superficial along the outer edge of the sartorius.

η.   The *vastus internus:* a very large muscle on the outer anterior aspect of the thigh; arising from the pelvis close to the hip-joint, it joins, below, two muscles on the back of the thigh (4. *a. β. γ*), and all end in a tendon which is inserted into the aponeurosis over the front of the os cruris.

e.   Cut across the belly of the sartorius, and turn its ends out of the way; dissect out the origin and insertion of the *adductor longus* and the *adductor magnus* (*d. ζ.* and *β*).

a.   The *adductor longus* arises from the anterior inferior part of the symphysis of the iliac bones; by its lower end it joins the adductor magnus.

β.   The *adductor magnus* arises from the pelvis, between the origin of the sartorius and that of the rectus internus major. Its fibres are inserted directly (*i.e.* without the intervention of a specialised tendon) into the inner side of the distal half of the femur.

3.   **The deep muscles on the front of the thigh.**

a.   Divide and reflect the adductor longus, rectus internus major and rectus internus minor. The following muscles will be displayed:—

a.   The *pectineus:* it lies at the upper part of the thigh immediately internal to the vastus internus: it arises from the front of the pelvis, close to the symphysis, and is inserted into the anterior surface of the distal half of the femur.

β.   The *adductor brevis* (2. *d.* γ): it lies along the inner side of the pectineus and arises and is inserted close to it.

γ.   The *semitendinosus:* this is a long slender muscle lying beneath the rectus internus major and bifurcated at its upper end: its two *heads*, thus formed, arise, one (*anterior head*) from the pelvis between the ischial symphysis and the acetabulum; the other (*posterior head*) from the ischial symphysis: the muscle terminates below in a rounded tendon which is inserted along with the sartorius.

4.   Turn the frog over on to its belly, and remove the skin from the back of the limb.

The muscles on the back of the thigh.   These are :—

a.   The *triceps femoris:* a very large muscle on the outer side, divided above into three heads, which are often regarded as separate muscles, viz.

a.   The *vastus internus:* the anterior division, which has been already seen on the front of the thigh.   (2. *d.* η.)

β.   The *vastus externus:* the posterior division ; it arises from the hinder edge of the iliac bone.

γ.   The *rectus femoris anticus:* the middle division of the triceps; it arises from the posterior part of the ventral border of the iliac bone.   For the insertion of the triceps femoris, see 2. *d. η.*

b.   The *glutæus:* this muscle arises from the hinder two-thirds of the external surface of the ilium; it runs down between the vastus externus and the rectus anticus to be inserted into the back of the head of the femur.

c.   The *pyriformis.*   This arises from the posterior part of the urostyle and, passing inside the vastus externus, is inserted into the shaft of the femur.

d.   The *biceps femoris:* a long thin muscle, lying along the inner side of the vastus externus; it arises from the iliac bone above the acetabulum; below it divides into two pieces, one of which is inserted into the middle of the shaft of the femur, while the other ends in a rounded tendon which is inserted into the back of the distal end of the same bone.

e.   The *semimembranosus:* a large muscle inside the biceps; it arises from the upper posterior part of the iliac symphysis and is inserted into the aponeurosis round the knee-joint.

Lying deep in the thigh, between the biceps and semimembranosus muscles, will be seen the *femoral vessels* and the *sciatic nerve.*

f.   Divide and reflect the vastus externus and the biceps: beneath them will be laid bare—

The *ileo-psoas*: it arises from the internal surface of the posterior part of the ilium and is inserted into the posterior aspect of the shaft of the femur.

g.   Remove the pyriformis, cut through the semi-membranosus close to its origin and throw it downwards; this brings into view a small triangular muscle—

The *quadratus femoris*, which arises from the ilium behind the acetabulum and is inserted into the middle of the shaft of the ventral surface of the femur.

h.   The *obturatorius*: this is a small muscle lying on the dorsal surface of the hip-joint.

5.   **The muscles of the leg.**

a.   Turn the frog on its back, and remove the skin from the foot: fix the dorsal surface of the foot upwards. The os cruris will now be seen running down the middle of the leg. Lying on its present inner (proper dorsal) side are two muscles, viz.—

a.   The *gastrocnemius*: a muscle with a great fleshy belly; it arises above by two tendons; one (much the larger) is inserted behind the knee-joint partly into the femur, partly into the os cruris; the other joins the aponeurosis on the outer side of the knee-joint. Below, the muscle ends in a great tendon (*tendo Achillis*), which terminates in an aponeurosis on the plantar surface of the foot.

β. The *tibialis posticus:* a slender muscle covered in great part by the gastrocnemius; it arises from the greater part of the posterior surface of the os cruris and, passing along the inner side of the ankle-joint, is inserted into the astragalus.

b. On the opposite side of the bone lie four muscles : viz.—

a. The *peroneus:* the largest and most external; it arises from the outer side of the distal articular end of the femur and, running past the outer side of the ankle-joint, is inserted into the calcaneum.

β. The *tibialis anticus:* a small muscle inside and beneath the peroneus; it arises from the front of the lower end of the femur and from the capsule of the knee-joint; below it divides into two parts, one inserted into the dorsal side of the astragalus, the other into the calcaneum.

γ. The *extensor cruris brevis:* this lies internal to the upper part of the last muscle; it arises from the front of the distal articular end of the femur and is inserted into the middle third of the os cruris.

δ. The *flexor tarsi anterior:* this arises where the last muscle ends, and is inserted into the dorsal side of the astragalus.

6. **The nerves of the hind-limb.**

These are now to be dissected out in that leg which has not been used for the muscles.

a. *The sciatic nerve.*

α. Find it on the dorsal side of the thigh by separating the biceps and semimembranosus muscles; it appears as a slender white cord.

β. Dissect it out carefully in the middle of the thigh, noting the branches it gives off to the various muscles.

γ. Follow it up to the abdominal cavity, cutting away the tissues lying between the ilium and urostyle, which cover it in.

δ. Follow it down towards the knee : a little way above the joint it divides into two branches; one (*posterior tibial*) runs inside the large head of the gastrocnemius muscle, the other (*peroneal*) between its two heads.

b. *The posterior tibial nerve.*

α. Note especially the branch which it gives off opposite the knee-joint, and which, after running for some way along the deep surface of the gastrocnemius, enters that muscle.

β. Follow the nerve down the leg : it runs along the tibialis posticus muscle, giving off branches here and there.

γ. Near the foot it turns to the back of the ankle-joint and enters the plantar surface of the foot, where it ends in a number of branches.

c. *The peroneal nerve.*

α. This runs down the leg close to the peroneus muscle, giving off branches on its way.

β.   Towards the bottom of the leg it comes to
the front of the ankle-joint, and divides into
branches which are distributed on the dorsum
of the foot.

[F.   DISSECTION OF THE VASCULAR SYSTEM.

The dissection of the blood-vessels is much facilitated by
previous injection: this may be done as follows: Dissolve
some gelatine, with the aid of heat, in a coloured fluid
(solution of carmine or of Berlin blue); the gelatine ought
to be added in such quantity that the fluid just sets when
cold: kill a frog with chloroform; lay bare its heart, taking
care not to injure the anterior abdominal vein; prick a
small hole in the venous sinus and sop up any blood that
flows out; pass a ligature round the *truncus arteriosus;*
make a small aperture in the ventricle and pass a glass
tube, drawn out to a fine point, through the ventricle into
the arterial bulb, and tie it in; fill this tube with normal
salt solution and connect it, by a bit of gutta percha tubing,
with a syringe filled with the injecting fluid, which should
not be warmer than 35° C.; inject very slowly and with
slight pressure. The venous system may also be readily
injected by dividing the anterior abdominal vein and
passing the syringe into its posterior end. When the injection is finished, put the animal in alcohol for some
hours.]

1.   **The anterior abdominal vein.**

a.   Carefully dissect the belly-walls away from the
anterior abdominal vein: at its posterior end
this vessel will be found to get a small branch
from the front of each thigh and then to divide
into two large trunks (*pelvic veins*) which run,
one on each side, towards the back of the thigh.

b.   Turn the animal over and follow one of these trunks back : it will be found to be continuous with the *sciatic vein*, which ends in the pelvis by dividing into this and another (*renal portau*) vessel.

c.   Trace the anterior abdominal vein forwards : it divides into two branches, one of which goes to the right and the other to the left lobe of the liver.

2.   Raise the liver, and note the **vena portæ** which enters its lower surface; it is formed by the union of a vein (*gastric*) from the stomach with one (*lieno-intestinal*) from the spleen and intestines.   The gastric division of the vena portæ communicates by a large branch with the left division of the anterior abdominal vein.

3.   **The veins of the head and neck and fore-limbs.**

a.   Remove the liver, being careful not to injure the inferior vena cava beneath it.

b.   Pass a bit of glass tube down the frog's gullet (in order to stretch out the neighbouring parts) and clean the aortic arches : passing in front of each aortic arch, near its point of division is—

c.   The *external jugular vein*, running up the side of the throat towards the angle of the lower jaw and receiving the veins of the mandibular and lingual regions.

d.   Follow this vein down towards the heart : a little way below the aortic arch it is joined by another large vein—

e.   The *subclavian :* follow this outwards; it will be found to be formed mainly by the union of two large branches : one (*axillary* or *brachial vein*) coming from the antebrachium and manus; the other (*musculo-cutaneous*) from the back and head.

f. The *innominate vein* is formed by the union of the *internal jugular vein*, which brings back the blood from the brain and spinal cord, with the *subscapular vein* returning the blood from the brachium and shoulder.

g. The *superior vena cava* (*right* and *left*): this is formed by the union of the subclavian, external jugular and innominate veins on each side : follow it to the heart, where it ends by entering the sinus venosus.

4. **The inferior vena cava and renal portal veins.**

a. Divide the alimentary canal above the stomach and also close to the cloaca, and remove the intermediate portion : dissect out the veins connected with the kidneys.

b. The *renal portal* vein : running from the bifurcation of the pelvic vein. to enter the lower-outer border of the kidney.

c. The *inferior vena cava :* the large vein lying between the kidneys and chiefly formed of branches from them, but also getting branches from the generative organs and the liver.

d. Follow it up to its anterior ending in the sinus venosus.

5. **The aortic arches and their branches.**

a. Dissect out the branches of the aortic arches: three on each side.

    a.　The anterior division (*carotid trunk*): it, after giving off a branch (*lingual artery*) which runs up the throat, ends in a small red body, *the carotid gland*, from which other arteries proceed.

    β.　*The systemic aortic arch:* this is the middle and largest division : it runs round the throat towards the vertebral column, giving off on its way the *subclavian artery* which runs to the fore-limb.

    γ.　*The pulmo-cutaneous artery*, or posterior division of the aortic arch : it runs to the root of the lung, giving off on its way a *cutaneous branch* which runs out to the integument about the shoulder.

    b.　Imbed in paraffin an aortic arch which has been hardened in spirit and cut transverse sections of it : examine with 1 inch obj.　Note the two partitions subdividing it into three channels.

6.　**The dorsal aorta and its branches.**

    a.　Remove the kidneys with vena cava inferior and the generative organs: the *dorsal aorta* is then laid bare lying on the bodies of the vertebræ.

    b.　Follow the *systemic aortæ* (5. *a.* β) round the neck; they will be found to unite beneath the vertebral column to form the dorsal aorta.

    c.　Follow the aorta backwards: it gives off many branches on its course; note the large one (*cœliaco-mesenteric artery*) arising from it just below its point of formation.

d. Small branches to the renal and generative
organs (only the cut ends of these branches can
now be found) and to the muscles of the back.

e. Near the pelvis it ends by dividing into two
trunks (the *iliac arteries*) which run behind the
pelvic bones, giving off hypogastric branches to
the bladder and the walls of the abdomen.

f. Turn the animal over on to its belly and trace
the iliac arteries backwards: they are mainly
continued down the thigh as the *femoral arteries.*

7. **The pulmonary veins.**

a. Trace them from the left auricle to the lungs.
Examine the left auricle carefully and find the
opening of the common pulmonary vein into it.

G. THE NERVOUS SYSTEM OF THE FROG.

1. *The method of exposing the brain and spinal cord.*—
Take a frog which has been a day or two in spirit;
divide the skin along the middle of the dorsal surface from the snout to the anus and reflect it to
each side, noting the small nerves running into it
on each side of the middle line; remove the muscles
lying on the arches of the vertebra; open the neural
canal by dividing the membrane between the atlas
and occiput; then introduce one blade of a small,
but strong, pair of scissors into the cranial cavity,
and cut away bit by bit the bones which form the
roof of the skull, taking care that the point of the
scissors does not injure the brain. Next remove
the upper part of the arches of the vertebræ in a
similar manner. A delicate pigmented membrane
(*the pia-mater*) is now laid bare, covering the brain;

over the spinal cord it is usually concealed by a
quantity of soft material; gently remove this with a
pair of forceps or wash it away with a syringe.

2. **The brain.**

On the dorsal aspect of the brain, which is now ex·
posed, the following parts will be seen;—

a. In front, two elongated masses forming about
the anterior half of the brain: a slight transverse depression divides each into an anterior
smaller and a posterior larger portion. The
inner faces of the anterior portions are closely
united together; those of the posterior portions
are separated by a cleft. The posterior portions are the *cerebral hemispheres* (*prosencepha-
lon*); the anterior, the bases of the *olfactory
lobes* (*rhinencephalon*).

a. The *olfactory lobes* become narrowed into two
rounded trunks (commonly termed the *olfactory
nerves*), which leave the skull, and applying
themselves to the outer face of the lining membrane of the nasal chamber, give off a number
of branches which are distributed on that membrane.

b. The *thalamencephalon:* lying between the posterior ends of the cerebral hemispheres: on it
are to be noticed—

α. The *pineal gland:* a very small mass in front;
not composed of nervous tissue.

β. The *thalami optici:* the nervous masses seen
below the pineal gland: between them lies a
narrow cavity, *the third ventricle.*

c.   The *optic . lobes* (*mesencephalon*): a pair of rounded eminences lying behind the thalamencephalon.

d.   The *cerebellum* (*metencephalon*): a narrow transverse band lying behind the optic lobes.

e.   The *medulla oblongata* (*myeloncephalon*): the portion of the brain lying behind the cerebellum.

    a.   Lying on the medulla oblongata is a triangular depression (4th *ventricle*) with its apex turned backwards.

3. **The spinal cord.**

a.   *Its form:* wide in front, but narrowing rapidly about the 5th or 6th vertebra, and thence continued along the neural canal as a slender tapering filament.

b.   The groove (*posterior fissure*) running along its middle line.

c.   The *spinal nerves* arising from it.

    a.   Ten on each side.

    β.   Each arising by two *roots* (*anterior* and *posterior*): these are most easily seen in the 7th, 8th and 9th nerves, where they are much longer than in the others. Sometimes the anterior root is double.

    γ.   The direction of the roots: directly outwards in anterior nerves; obliquely backwards in the 4th, 5th and 6th; almost directly backwards for a considerable distance in the neural canal in the 7th, 8th, 9th and 10th.

δ. The point of union of the roots to form a nerve trunk, in the intervertebral foramina.

Draw the exposed parts of the brain and spinal cord.

4. Divide the olfactory lobes, and raise the front end of the brain; turning it back gradually, divide with a sharp scalpel any nerves that are seen running from it to the cranial walls: most of the nerves being small, they will probably be torn across unobserved, but the large optic nerves will at any rate be seen: next divide the nerve-roots of the spinal cord; remove it and the brain together and place them with the ventral side upwards.

 a. On the under surface (*base*) of the brain will be seen—

  α. *The optic commissure* or *chiasma* opposite the posterior end of the cerebral hemispheres; with the *optic nerves* diverging from its anterior end and the *optic tracts* entering it posteriorly.

  β. Lying behind the optic commissure, between the optic tracts, is a small eminence—the *pituitary body*.

  γ. On each side of the two last-mentioned structures and arched over in front by the optic tracts, are the *crura cerebri.*

 b. Divide the cerebral hemispheres horizontally with a sharp scalpel: in each will be found a cavity— the *lateral ventricle.* Each lateral ventricle communicates with the third ventricle. The optic lobes, divided in the same way, will be seen to roof over a cavity which communicates with the third ventricle in front, and with the fourth ventricle behind.

c.   *The spinal cord.*

   α.   The *anterior fissure* running along its ventral surface.

   β.   Its form: subcylindrical; wider from side to side (especially opposite the second pair of nerves) than dorso-ventrally.

   γ.   Imbed it in paraffin and cut a transverse section: mount in glycerine, and examine with 1 inch obj.; note its peripheral portion (*white matter*) different in appearance from the central parts (*grey matter*): the canal (*canalis centralis*) running up its centre.

5   Turn the frog over on to its back, lay open its abdominal cavity and remove all the alimentary canal from the gullet to the rectum, along with the liver, kidneys and generative organs.

6.   **The sciatic plexus.**

   a.   This is now seen as a number of large nerve-cords on each side of the dorsal aorta; note the communications between the different cords of the same side.

   b.   Follow down the plexus on one side: it ends below in a large trunk which is continuous with the sciatic nerve.

   c.   Trace the nerve-trunks forming the plexus up to the spinal column. They are continuous with the 7th, 8th, and 9th spinal nerves.

7. In front of the sciatic plexus—lying on the muscles bounding posteriorly the abdominal cavity—are three nerves running obliquely downwards and outwards on each side: they are continuous with the 4th, 5th and 6th spinal roots.

8. **Some nerves of the neck.**

   a. Pass a piece of tubing down the gullet so as to distend it: and then carefully remove the mylo-hyoid muscle (B. 6. *a*).

   b. Find the posterior cornu of the hyoid bone on one side: from it a slip of muscle (*petrohyoia*) will be seen passing up towards the occipital region of the skull. Lying along the posterior border of this muscle is the *pneumogastric nerve;* follow its branch to the heart.

   c. Lying on the petrohyoid and in front of the pneumogastric, from which it arises, is the *laryngeal nerve.*

   d. Some way in front of the laryngeal nerve is seen the *glossopharyngeal nerve,* turning up towards the front of the jaw.

   e. Superficial to the glossopharyngeal, but with a generally similar direction, is the *hypoglossal nerve* (B. 6. *a*).

9. **The brachial nerve.**

   Lay it bare in the arm-pit and follow it back to the spinal cord: it is formed by the union of the 2nd and 3rd spinal nerves.

10. **The sympathetic system.**

    a. Gently raise the aorta: along each side of it will be found the main *sympathetic trunk.* A slender

cord, with enlargements (*ganglia*) on it at intervals.

b.   Note the branches passing between its ganglia
and the nerves of the sciatic plexus.

c.   Carefully dissect out the gangliated cord for its
whole length : ten ganglia, each provided with
communicating branches to other (*spinal*) nerves,
will be found on it.

H.   THE ORGANS OF SPECIAL SENSE.

The complete examination of these, especially as re-
gards their histology, is difficult, and necessitates the
employment of niceties in manipulation which it would
be beyond the scope of this work to describe, so that in
the following account attention is mainly given to those
points which can be made out without the microscope.
A brief account of the microscopic structure of the retina will however be found below (J. **h**).

a.   **The Eye.**

1.   Take an uninjured frog and examine its eye.   It will
normally be found to project considerably above the
top of the head, but if touched it is withdrawn into
a sort of socket.   If the animal's mouth be opened,
an elevation, caused by the eye-ball, will be seen on
its roof, and this is more prominent when the eye-ball
is retracted.

a.   Gently touch the eye and observe how it is closed,
by the pulling over it of the lower transparent
eye-lid.   The upper eye-lid is very small and
hardly moveable.

    b. When the eye is open, observe the parts exposed.

        a. The transparent *cornea* covering all its exposed surface.

        β. Through the cornea is seen the *iris*, a membrane coloured by brown and golden pigment, the latter forming a very brilliant ring around the inner margin of the iris. The lower margin of this ring is interrupted at one point, the yellow pigment being there absent, and from the break a faint dark line can be traced downwards through the rest of the lower part of the iris.

        γ. The elliptical opening or *pupil* in the middle of the iris with its long axis directed antero-posteriorly.

  2. Kill the frog (by chloroform or by pithing), and carefully dissect away the parts from around the eye-ball, cutting away, with the rest, the part of the upper jawbone which forms the lower boundary of the socket of the eye-ball, or the *orbital cavity.*

        a. As the surrounding tissues are cleared away from the eye-ball, notice the small muscles which are inserted into it.

        β. At the back of the eye-ball and passing into it will be found the *optic nerve.*

  3. Divide the optic nerve and, having thus detached the eye, pin it to a piece of loaded cork, with the corneal surface upwards.

        a. Notice the more opaque coat (*sclerotic*), with which the margin of the cornea is continuous, and which forms the outer envelope of the eye-ball on its sides and back. In some parts the sclerotic is semi-transparent and allows the pig-mented choroid coat (3. *g*) to be more or less distinctly seen through it.

b. Prick the cornea with the point of a sharp scalpel, taking care not to injure the iris; note the clear *aqueous humour* which spirts out, the cornea at the same time collapsing.

c. Seize the cut edge of the cornea with a fine pair of forceps and, with sharp scissors, carefully cut through it all round at the line of junction with the sclerotic. The convex anterior surface of the transparent *crystalline lens* will now be seen pro-jecting through the pupil.

d. Place the cork in a vessel of convenient size and add enough water to cover the eye. Then, with sharp scissors, cut away the iris, and so expose all the anterior surface of the lens. Passing the point of a scalpel under one edge of the lens, gently tilt it out and examine it.

   a. The *crystalline lens* of the frog is nearly sphe-rical but somewhat thicker from side to side than antero-posteriorly. Its anterior surface (that which projects into the pupil) is also less convex than its posterior surface.

e. The cavity of the *posterior chamber* of the eye is now exposed. It is filled with a gelatinous transparent mass, the *vitreous humour*, which can be seen if the water, in which the eye is being dissected, be poured away.

f.   Lining the posterior chamber of the eye is the *retina*, which, from the action of the water, will probably now be somewhat cloudy: in the unaltered state it is perfectly transparent, allowing the choroid (3. *g*) to be seen through it. With the point of a microscope needle gently raise the retina from the black membrane (choroid) beneath it.   It will be found that this can be readily done except at one point (answering to the *blind spot* of our own eyes), which, by turning the eye-ball over will be seen to be opposite the point of entrance of the optic nerve (2. *β*).

g.   The *choroid coat* of the eye-ball is now exposed. It is a dark pigmented membrane, of loose flocculent texture, which can be readily detached, by needles, from the sclerotic which lies outside it.

b. The Ear.

1.   The frog has no external ear, its tympanic membrane, as already mentioned (A. 2. **a**), being exposed on each side of the head.

a.   Note the arrangement of the tympanic membrane : it is smoothly stretched over a hard ring.

b.   Dissect away the outer or integumentary layer of the tympanic membrane.   Beneath it will be found a transparent membrane, formed by the fibrous and mucous layers, with an opaque white patch about its middle.

c.   Cut through these layers of the tympanic membrane along their margins : the tympanic cavity will then be laid open.

α. The tympanum of the frog is a funnel-shaped cavity with its wider end turned outwards. Its sides are bounded by a smooth slightly pigmented mucous membrane, continuous with that of the mouth through the Eustachian recess.

β. In its roof lies a rod, ossified in the middle, cartilaginous at each end, which is the *columella auris.* The columella is attached, by its inner end, to the upper and anterior part of the inner wall of the tympanum and, by its outer end, to the middle layer of the tympanic membrane, in the region of the opaque patch mentioned above (1. *b*).

γ. Close to the inner attachment of the columella there is a comparatively large oval opening in the wall of the tympanic cavity; this is the outer end of the Eustachian recess, the inner end of which has been already (B. 11. *a*) seen on the posterior part of the roof of the mouth. Pass a probe through the opening now exposed, and open the frog's mouth to see its passage into that cavity.

2. *The internal ear.*

a. Carefully dissect away the columella auris from its inner attachment. An aperture into which it was inserted will thus be exposed : this is the *fenestra ovalis.*

b. Take a pair of scissors and cut through the bones of the side of the skull in a line joining the fenestra ovalis and the "guard" of the parasphenoid bone (D. c. 3. *a*); the chamber in the pro-otic bone (D. c. 1. *d*) which is thus exposed contains part of the internal ear.

c.   The dissection of the internal ear of the frog is difficult, on account of its minuteness. But by carefully removing the bony and cartilaginous walls of the periotic capsule, piecemeal, the semicircular canals will be exposed and the whole membranous labyrinth may be extracted. It should be placed in a watch-glass containing salt solution or spirit, and its form studied under the simple microscope.

c.   **The . olfactory organs.**

1.   These consist of two chambers which open externally, near the end of the snout, by the *anterior nares*, and posteriorly into the mouth, just behind the vomerine teeth by the *posterior nares*.   Make out these openings.

a.   Take a frog which has been preserved in spirit and pass the point of a small pair of scissors into the external nostril of one side and cut away the roof of the nasal cavity.   A chamber is thus exposed which has a somewhat triangular form, the apex of the triangle being at the external nostril and the posterior nostril being at another angle and farther from the middle line.

b.   The walls of the cavity are slightly folded, and there is a well-marked hemispherical eminence on its floor.

c.   Open the other nasal cavity in a similar way: notice the boundary wall (*septum narium*) which lies between the two and completely separates them.

d. Open the nasal cavity of a frog which has been preserved in Müller's fluid ; gently scrape away a little of the epithelium lining the chamber, mount in water: examine with your highest power.

    a. Among numerous mutilated cells, a certain number of more or less perfect ones will be found : these are of two kinds, viz. large columnar epithelial cells (J. 1. *b*), each with an oval nucleus, an unbranched peripheral process and a branched deeper one ; and smaller cells, with less protoplasm around the nucleus and finer peripheral and central processes.

d. The gustatory organ.

1. The shape and arrangement of the frog's tongue have already been described (B. 11. *a*).

    a. Snip off a bit of mucous membrane from the upper surface of the tongue of a recently killed frog, mount in normal saline solution and cover in plenty of the fluid with a large coverslip : examine with one inch obj.

        a. On the surface of the fragment and especially· around its edges numerous minute elevations of the surface will be seen : these are the *papillæ:* some (*filiform papillæ*) are pointed at the free end and others (*fungiform papillæ*) flattened. Note the loops which the blood capillaries make in some of the papillæ.

        β. Examine one of the thinner bits of the specimen with a higher power : the papillæ will be seen to be covered by epithelium, which is for the most part ciliated (J. 1. *c*); some of the papillæ however will be seen to have no cilia except a narrow belt around the somewhat truncated apex; it is on these latter papillæ that the *gustatory discs* are placed, and in fortunate specimens nerve-fibres can be seen entering them.

J. SOME OF THE MORE IMPORTANT POINTS IN THE HISTO-LOGY OF THE FROG.

a. Epithelium.

1. This consists of cells which line free surfaces within the body: the epidermis covering the skin is a similar structure, and is continuous with epithelium at the apertures of the body. There are several main types of epithelium, viz.—

    a. Scaly epithelium. Open the abdomen of a recently killed frog, carefully remove the viscera and lay bare the lymph sinus at the back of the body-cavity. Cut away its thin wall as carefully as possible, taking great care not to drag or pull it. Place the fragment in $0.5\frac{0}{0}$ solution of silver nitrate for about three minutes: then remove, wash well in distilled water, and finally leave the specimen in distilled water and exposed to the sunlight. So soon as the bit of pleuroperitoneum has become of a well-marked brown colour, mount it in glycerine and examine with a high power.

    a. It will be seen to be covered on both sides with flat closely fitting cells, the boundary lines of which are stained black by the silver; according to the amount of staining a nucleus may or may not be rendered conspicuous in each cell.

Here and there in good specimens, rings of smaller and more deeply stained cells will be seen surrounding minute apertures (*stomata*).

b. *Columnar epithelium.* Scrape gently the inner surface of the mucous membrane of the intestine of a frog which has been preserved in Müller's fluid; mount the detached fragments in water and examine with a high power.

α. Numerous elongated cells, flat at one end and somewhat pointed at the other, will be seen. Each has a well-marked oval nucleus.

β. These cells may be seen *in situ* if a thin section of the hardened mucous membrane of the intestine or stomach be examined. They are closely applied and arranged in a single layer.

c. *Ciliated epithelium.* Snip off a bit of the mucous membrane from the tongue of a recently killed frog with a sharp pair of scissors: mount the bit in 0·75$\frac{9}{6}$ sol. of sodic chloride, avoiding pressure, and examine with a high microscopic power.

α. Note the shimmering appearance along its free edge, produced by the rapidly moving cilia; as the cilia begin to die and their movement slackens, individual ones can be seen.

β. Scrape gently, with a scalpel, one of the prominences in the roof of the frog's mouth beneath the eye-balls: mount the scrapings in
o·75 salt solution and examine with ⅛ objective for individual ciliated cells; note their
roundish form, granular protoplasm and nucleus, and the group of cilia borne on one
end; stain with iodine.

b. Cartilage.

1. Dissect out carefully the omosternal or xiphisternal
cartilage of a recently killed frog; mount in o·75%
sodic chloride solution and examine with ¼ or ⅛ obj.

a, Large roundish granular *cartilage-cells* will
be seen imbedded in a structureless or very
finely granular *matrix*, which is more refractive next the cells than elsewhere, and causes
the appearance of a sort of halo round each.

β. In each cell lies a distinct granular round
nucleus or sometimes two nuclei, containing
large highly refractive molecules.

γ. If the preparation be carefully made, each
cell will at first completely fill the cavity of
the matrix in which it lies, but if it be kept
some time or be treated with distilled water
the cells contract and so leave a transparent
ring between their surface and the inside of
the cavity in which each lies.

c. Bone.

a. Examine a prepared transverse section of hard
bone (say a bit of the shaft of the humerus or
femur) with 1 inch obj. A bit of mammalian

bone is the best for seeing the essential structure.

α. *The Haversian canals:* round or oval spaces, usually filled with dirt in grinding down the bone, and therefore black and opaque, but sometimes clear and empty.

β. *The lamellæ:* a series of concentric layers round each Haversian canal.

γ. *The lacunæ:* oval black spots between the lamellæ.

δ. *The canaliculi:* minute black lines seen radiating from the lacunæ.

ε. Besides the lamellæ above mentioned others will be seen which belong to no Haversian system, but either fit in the angles between the systems, or run around the outer surface of the bone.

b. Examine with $\frac{1}{8}$ obj. Observe the lacunæ and canaliculi more accurately.

c. Examine in water or glycerine a thin transverse section of a long bone which has been softened by dilute acid.

α. The *Haversian canals* empty or containing a granular mass.

β. The *lamellæ:* very indistinct.

γ. The *lacunæ* seen as transparent oval spaces.

δ. The *canaliculi* transparent and almost invisible.

d. Examine sections of softened bone which have been stained with carmine: in each lacuna will be found a stained mass of protoplasm.

c. Examine longitudinal sections of the femur or humerus: the *Haversian canals* are seen to be channels running for the most part in the long axis of the bone, but communicating with one another frequently by cross branches. The *lacunæ*, &c. appear much as in the transverse section.

d. Connective tissue.

1. Of this there are two main varieties, viz.—

a. White fibrous tissue. This occurs nearly pure in tendons, but is widely distributed throughout the body, mixed with other tissues. Tease out a bit of fresh tendon in water: examine with a high power.

a. It is chiefly made up of very fine wavy fibres which run in bundles parallel to one another; they have an ill-defined outline and do not branch.

β. Treat with dilute acetic acid. Most of the fibres disappear, but a few well-defined curled fibres (*yellow elastic fibres*, see *b*) remain. Besides these some elongated granular protoplasmic masses are brought into view (*connective-tissue corpuscles*).

b. Yellow elastic tissue. This does not occur in large collections, in the pure form, in the frog; although mixed with white fibrous and other tissues, it is very widely distributed.

a. Tease out in acetic acid some of the bands of tissue beneath the frog's skin; examine with a high power. Numerous fine well-defined branched fibres, running in bold curves, will be seen: these are yellow elastic fibres, as the acetic acid (*a. β*) destroys the white fibrous tissue.

e. Striped muscle.

a. Tease out gently a bit of muscle which has been kept in alcohol, and examine with 1 inch obj.

  a. Composed of elongated *fibres*, which exhibit a tendency to split up into finer filaments (*fibrillæ*).

b. Examine with high power.

  a. The alternate lighter and darker bands placed transversely to the long axis of the fibre (*transverse striation*).

  β. The fine structureless membrane (*sarcolemma*) enveloping the fibre: seen here and there as a delicate film where the fibre is twisted or bruised.

  γ. The tendency to split up into fibrillæ.

c. Tease out a bit of fresh muscle in $0.75\frac{0}{0}$ sodic chloride solution.

  a. The transverse striation of the fibres: less distinct than in the muscle from alcohol.

  β. The absence of a tendency to split up into fibrillæ.

  γ. The sarcolemma, which may be seen at points where the continuity of its contents has been interrupted by pressure, twisting, &c.

δ. Treat with acetic acid: the striation rendered very indistinct; oval nuclei made apparent here and there in the fibre.

f. Nerve-fibres.

1. Tease out a bit of fresh nerve in 0·75⅔% sodic chloride. Examine with a high power.

   a. Composed of well-defined fibres (*white nerve fibres*) mixed with white fibrous tissue (**d.** 1. *a*).

   b. The *appearance* of the nerve-fibres: each has a double contour, marked out by a curdled-looking highly refractive border on each side.

   c. *The structure of the nerve-fibres.*

      α. The delicate structureless investing membrane (*primitive sheath*); look for it on bruised nerve-fibres or at the ends of torn fibres.

      β. The highly refractive border (*medullary sheath*) within the primitive sheath.

      γ. The central homogeneous axis (*axis cylinder*); look for it projecting beyond the medullary sheath of torn fibres.

   d. Treat a fresh bit of teased-out nerve with chloroform: the medullary sheath will be dissolved out, and the axis cylinder plainly seen.

g. Nerve-cells.

1. Take a sympathetic ganglion from a recently killed frog: dissect out in normal saline solution and examine with ⅛ obj.

   a. Among the pigment-cells, which are clustered around the ganglion, will be seen numerous large pale granular cells, each possessing a conspicuous clear round nucleus with a distinct nucleolus.

β. Tease out in glycerine a Gasserian ganglion from a frog's head which has been preserved in Müller's fluid or chromic acid. Cells like those described above will be found.

γ. Examine sections of spinal cord which have been stained with carmine or hæmatoxylin, and note the large branched nucleated cells in the grey matter especially towards the ventral side of the cord.

h. The retina.

ı. Sufficiently satisfactory specimens of this organ can be obtained as follows. Take perfectly fresh eyes from a frog, prick the corneas in two or three places, and lay the eyes aside for three or four days in $0\cdot25\frac{0}{0}$ chromic acid solution: then transfer them to alcohol and keep them in it until wanted.

a. Carefully cut open an eye preserved in the above method and expose the retina: transfer the latter to a glass slide, and with a razor chop down on it so as to cut off a number of slices: add glycerine, put on a cover, and examine with a low power. Some of the bits will be found thin enough for further examination.

b. With the low power little can be seen but that the retina is composed of a number of different layers, some of which appear less opaque than the others.

    c. Put on a high power and examine, make out the following points—

      a. *The internal limiting membrane,* a thin structureless layer.

      β. *The nerve-fibre layer:* thin and granular.

        [Both *a* and *β* are often difficult to make out in retinas prepared as above.]

      γ. *The nerve-cell layer:* composed mainly of cells like those described above (**g.** 1. *a*), but rather smaller than those from the sympathetic ganglia. From some, branches can be traced into the next layer.

      δ. *The molecular layer:* this is thicker than any of the preceding, and has a finely punctated appearance: running through it the fibres of Müller (**h.** 1. *ι*) are very plainly seen.

      ε. *The inner granular layer:* this is the layer which usually looks clearest in sections, its elements being less closely packed than those of the other layers. It is made up of a number of nuclei (which in chromic-acid specimens look granular), around which is collected a very small amount of protoplasm, and of fine fibres, some of which can be traced joining the nuclei or *granules.*

      ζ. *The inter-granular (fenestrated) layer.* A narrow cloudy layer in which no definite structural elements are visible.

      η. *The outer granular layer.* Much thinner than the inner granular layer and more closely packed. It is composed of distinct fibres (*rod-* and *cone-fibres*), each of which swells out and has a nucleus (the *granule*) developed in the enlargement.

θ. *The external limiting membrane.* A thin homogeneous layer like α.

ι. *The fibres of Müller.* These are highly refracting fibres which can be traced with ease from the internal limiting membrane to the fenestrated layer. They probably run beyond the latter and end on the external limiting membrane, but are difficult to trace through the outer granular layer.

κ. *The rod- and cone-layer.* The main thing which will be noted here is the huge *rods* for the most part distorted by the treatment to which the retina has been exposed. In favourable bits it can be seen that each rod is divided transversely into an inner and an outer segment. The *cones* are few and small, and generally completely concealed by the rods.

d. Take a fresh frog's eye: prick its cornea and collect the aqueous humour on a slide. Then open the eye, remove a bit of the retina and tease it out in the aqueous humour, mount and examine with a high power.

α. Numerous rods will be seen floating about, many broken but some intact and shewing the boundary line between their two segments very plainly. At first both segments are homogeneous, but very soon they begin to alter; the

outer layer frequently then getting a transversely striated appearance and shewing a
tendency to split up into corresponding pieces :
gradually these rods entirely disintegrate, first
curling up, swelling out, &c.

i. The skin.

1. Cut out a piece of skin from the back of the thigh of
    a recently killed frog : spread it out in water, cover,
    and examine with a low power : note—

    a. *The pigment-cells;* seen as black irregularly
        shaped patches; some compact, others more or
        less branched.

    b. *The mouths of the cutaneous glands;* seen as
        clear round spots, although their openings are
        really triradiate : their number.

2. Take a piece of skin that has lain for a day or two in
    solution of ammonia bichromate and then in alcohol:
    imbed it, and cut sections perpendicular to its sur-
    faces: mount in glycerine. Examine with a low
    power ; note—

    a. The two layers of the skin, *dermis* and *epidermis,*
        the former being much the thicker : note in the
        dermis its deeper connective-tissue layer, and
        its more superficial granular layer immediately
        beneath the epidermis.

    b. Examine with a higher power.

        a. The *epidermis* is seen to be made up of nume-
            rous closely packed cells, arranged in several
            layers.

        β. The deepest epidermic cells are granular, nucleated, and somewhat oval, with their long axes at right angles to the surface.

γ. Then come several rows of cells, also granular and nucleated, but becoming smaller and rounder as they become more superficial.

δ. The most superficial three or four layers of cells are flattened parallel to the surface, are not granular, and possess no apparent nucleus.

ε. Here and there a pigment-cell is seen among the epidermic cells, and some of the latter contain a few pigment-granules.

ζ. *The dermis*, consisting fundamentally of white fibrous and elastic tissues: its glandular and non-glandular layers.

η. Immediately beneath the epidermis is a thin stratum of connective tissue in which lie many large pigment-cells, sometimes forming an almost continuous layer.

θ. Then come a large number of round cavities, the *cutaneous glands*, lined by large, pale, slightly granular, nucleated cells, which are columnar when seen sidewise, but polygonal when seen from the base or apex. Sometimes the duct of the gland can be traced running from it through the epidermis. Separating the glands and supporting the epithelium are bundles of connective tissue, consisting mainly of fibres running perpendicularly to the surface.

ι. The deepest layer of the dermis is made up

of connective-tissue bundles, running for the most part parallel to the surface.

j. The kidney.

1. Take a frog's kidney which has been for a week in solution of bichromate of potash, and then for a day or two in spirit. Imbed it, cut sections parallel to its flatter surfaces, and mount in glycerine.

   a. Examine with a low power.

      α. Note the numerous *tubules* of which the organ is mainly composed and which twist about in all directions, and are consequently cut, some transversely, some obliquely, and others more or less longitudinally. The absence of any marked division into cortex and medulla.

      β. The clear round holes, scattered about; these are sections of *glomeruli* from which the contained vessels have fallen out. Some may be seen in which a granular mass still lies.

   b. Examine with a higher power—

      α. The *epithelium* lining the tubules, composed in some of granular and ill-defined cells, in other (usually larger) tubules of clearer and better-defined cells; both varieties are nucleated.

   c. Examine specimens of injected kidney with a low power. Note the vascular tufts of the glomeruli.

k. The testis.

1. Imbed a testis which has been hardened in alcohol: cut sections and mount in glycerine.

a.   Examine with a low power.

    α.   The organ is chiefly made up of tortuous tubules, which are seen cut in various directions.

b.   Examine with a high power.

    α.   Note the epithelium lining the tubules : it varies with the season of the year (whether before or after the breeding-time), and is usually extremely granular and ill-defined. The cells are arranged in two or three rows, and at the time of breeding the most superficial layer of cells is transformed into spermatozoa, each cell giving rise to several. These lie side by side at right angles to the lumen of the tubule, which accordingly appears to be lined by them.

c.   The *spermatozoa* (B. 10. *a.* γ).

1. The ovary.

1.   The structure of this organ is easiest made out shortly after the breeding-time. Remove one of the ovaries, place it in water, and make an incision into it : it will be seen to contain a cavity, and projecting upon the walls of this cavity and also upon the outer surface of the ovary are numerous round eminences of various sizes : these are *ova* in different stages of development, and the large ones will be seen to have become more or less pigmented.

2.   Tease out a bit of ovary in normal saline solution : cover, and examine with a low power.

    a.   Note the ova, many much smaller than those which were seen (1) with the naked eye : they appear as granular spherical masses with a clearer central patch.

b.   Examine with a high power a portion of your specimen containing some of the younger and more transparent ova.   Note—

   α.   The thin structureless membrane, *vitelline membrane*, enveloping each.

   β.   The granular matter (*yelk, vitellus*) forming most of the ovum.   It sometimes appears to be composed of an outer granular and an inner clearer layer.

   γ.   The clearer central mass (*germinal vesicle*) imbedded in the vitellus.   The large number of highly refracting masses (*germinal spots*) within the germinal vesicle.

K.   The physiological properties of muscle and nerve.

Place a frog under a beaker, with a drop or two of chloroform : take it out immediately it becomes unconscious, which will probably be in a few seconds.   Now feel with a finger-nail for the depression beneath the skin at the back of the animal's head, which indicates the point of articulation of skull and spinal column : it lies in a line joining the posterior borders of the two tympanic membranes.   Divide the skin and muscles at this point until the neural canal is laid open, and then pass a stout wire into the cranium and down the neural canal of the vertebral column.   By this process (known as *pithing*) the frog is rendered totally incapable of further consciousness, though most of its tissues will retain their vitality for some time.

a.   Remove the skin from one leg, so as to lay bare
     the muscles : send an interrupted electric cur-
     rent through any one of them (or tap the muscle
     sharply with the back of a scalpel) : it will im-
     mediately *contract*, or *alter its form in a definite
     way;* it becomes *shorter and thicker*, and in so
     doing moves the bones to which it is attached.

b.   Very carefully lay bare the sciatic nerve, taking
     care not to crush or drag it : divide it as high
     up as possible and, seizing it with a pair of for-
     ceps close to its cut end, lay it over the elec-
     trodes of an induction-coil. Probably when the
     nerve is cut the muscles of the limb will con-
     tract : whether or not, however, they will con-
     tract violently while the interrupted current is
     going through the nerve.

     [If an induction-coil is not at hand a bit of
     clean copper wire twisted round a strip of zinc,
     with the points of contact moistened with dilute
     acetic acid, may be used to stimulate the nerve;
     smart tapping or pinching with a pair of forceps
     will also excite it, but by such means the nerve is
     soon killed.]

The above experiments shew :—

c.   That the muscle is *irritable* and *contractile :*
     certain external agencies (*stimuli*) excite some
     change in it, the result of which is a muscular
     contraction.

d.   The nerve is *irritable :* certain external agencies
     excite some change in it, which in this par-
     ticular case manifests itself by a contraction of
     the muscles connected with the nerve.

    e.   The nerve possesses *conductivity :* although it is stimulated at some distance from the muscles, yet the change excited by the stimulus travels along it to them.

APPENDIX.

The various re-agents, mentioned in the "Laboratory work" in the preceding pages, are prepared as follows :

1. **Acetic acid, Dilute.**

   Mix 1 cub. centimetre of glacial acetic acid with 99 cub. cent. of distilled water.

2. **Ammonic bichromate, Solution of.**

   Dissolve 10 grammes of crystallized ammonic bichromate in a litre of distilled water.

3. **Carmine, Solution of.**

   Carmine.............................. 2 grammes.
   Strong solution of ammonia ...... 4 cub. cent.
   Distilled water .... 48 cub. cent.

   Dissolve the carmine in the ammonia and water; leave in an unstoppered bottle until nearly all smell of ammonia has gone. Afterwards keep in a well-closed bottle. Dilute a small quantity with fifteen or twenty times its bulk of water, when required for use.

4. **Chromic acid, Solution of.**

   Dissolve 10 grammes of crystals of chromic acid in one litre of water. This gives a 1 per cent. solution, from which weaker ones can readily be prepared. when required.

5. **Hæmatoxylin, Solution of.**

 a. Prepare a saturated solution of crystallized calcic chloride in 70 per cent. alcohol; then add alum to saturation.

 b. Prepare a saturated solution of alum in 70 per cent. alcohol. Add 1 volume of *a* to 8 of *b.*

 c. To the mixture of *a* and *b* add a few drops of a saturated solution of pure hæmatoxylin in absolute alcohol. Filter.

6. **Iodine, Solution of.**

 Prepare a saturated solution of potassic iodide in distilled water; saturate this solution with iodine. Filter. Dilute to a brown sherry colour.

7. **Magenta, Solution of.**

 Dissolve 1 decigr. of crystallized magenta (roseine) in 160 cubic centimetres of distilled water: add 1 cub. cent. of absolute alcohol. Keep in a well-closed bottle.

8. **Mayer's Solution.**

 See note p. 8.

9. **Müller's Solution.**
 Bichromate of potash .... 25 grammes.
 Sodic sulphate .... 10 grammes.
 Distilled water .... 1 litre.

10. **Osmic Acid, Solution of.**

 Best bought ready made in the form of 1 per cent. solution.

11. **Paraffin.**

 Melt together one part of solid paraffin (paraffin candles will do), one part of paraffin oil and one part of pig's lard. A mixture in the above proportions gives, when it has cooled, a mass of the most generally useful consistency.

To imbed an object, scoop a hole in a bit of the paraffin, place the object (the surface of which must be dry) in this hole and fill up the latter with some melted paraffin.

12. **Pasteur's Solution.**

See note, p. 6.

13. **Potash Solution.**

Dissolve 5 grammes of potassic hydrate in 100 cubic cent. of water.

14. **Schultz's Solution.**

Dissolve some zinc in hydrochloric acid; permit the solution to evaporate, in contact with metallic zinc until it has attained a syrupy consistence. Saturate the syrup with potassic iodide, and then add enough iodine to make a dark sherry-coloured solution. The object to be stained must be placed in a little water, and then some of the above solution added.

15. **Silver Nitrate, Solution of.**

Dissolve 0·5 grammes of silver nitrate in 100 cubic cent. of distilled water. Keep in an opaque stoppered bottle.

16. **Sodic Chloride, Solution of.** (*Normal saline solution. Salt solution.*)

Dissolve 7·5 grammes of sodic chloride in 1 litre of distilled water.

INDEX.

A.

ABDUCENTES, nervi, 189
Acetabulum, 224
Acrogenous growth, 47
Adductor muscles, 108, 116, 123
Alæ, 84
Alcoholic fermentation, 5, 9, 10
Algæ, 48
Alimentary canal, of Anodonta, 110, 121; of Crayfish, 131, 148; of Frog, 167, 173, 205; of Lobster, 131, 148; of Tadpole, 163
Alinasal process, 171
Alternation of generations, 37, 47, 61
Ambulatory limbs, 129, 151
Amœba, 17; Laboratory work, 21
Amœboid movements, 20, 105
Anacharis, protoplasmic movements in, 54
Angulo-splenial, 220
Annulus, 66
Anodonta cygnæa, 107; Laboratory work, 113
Antennæ, 130, 153
Antennules, 130, 153
Anterior commissure, 186
Anterior abdominal vein, 180, 200, 234
Anther, 70, 84
Antheridium, 43, 45, 51, 60, 68
Antherozooids, 46, 52, 61, 68
Aortic arches, 176, 203, 236
Appendages, of Bean, 70, 78; of Chara, 42, 48; of Crayfish, 128, 142, 151; of Frog, 160, 197, 212; of Lobster, 128, 140, 151
Aqueous humour, 246

Arachnoid membrane, 168
Archegonia, 61, 68
Arterial system, of Anodonta, 111; of Crayfish, 133, 145, 149; of Frog, 177, 237; of Lobster, 133, 145, 149
Artery, cœliac, 178, cœliaco-mesenteric, 178, 237; cutaneous, 178; femoral, 178, 238; hypogastric, 178, 238; iliac, 178, 238; lingual, 177, 237; mesenteric, 178; œsophageal, 178; pulmocutaneous, 178, 203, 237; pulmonary, 176, 178; subclavian, 178; vertebral, 178
Articular process, 213
Arytenoid cartilages, 181
Asci, 36
Ascospores, 8, 36, 41
Astacus fluviatilis, 127; Laboratory work, 140
Astragalus, 225
Atlas vertebra, 213
Atrium, 175, 201
Auditorii, nervi, 190
Auditory organs, of Anodonta, 113, 121; of Crayfish, 138, 156; of Frog, 194, 247; of Lobster, 138, 156
Axillary vein, 236
Axis cylinder, 257

B.

Bacillus, 28
Bacteria, 25; Laboratory work, 27
Balantidium, 93
Bark, 72
Basipodite, 151

Crura cerebri, 186, 241
Crural nerve, 193
Crystalline lens, 137, 154, 246
Cutaneous glands, 262; artery, 178

D.

DACTYLOPODITE, 151
Dentary bone, 220
Dermis, 261
Development, of Anodonta, 112; of Bean, 71; of Chara, 46, 50; of Crayfish, 139; of Fern, 60; of Frog, 162; of Lobster, 139; of Mucor, 34, 37; of Penicillium, 32, 40
Diaphragm, 166
Dorsal aorta, 178, 237
Dorso-lumbar vein, 180
Dotted ducts, 81
Duodenum, 173

E.

EAR, see Auditory organ
Ecdysis, 139
Ectoderm, 100, 103
Ectosarc, 18, 21
Embryo, of Anodonta, 112, 126; of Bean, 71, 86; of Chara, 47; of Fern, 61; of Frog, 162; of Lobster, 139
Embryo, cell 61, 71, 86; sac, 71, 86, 88
Encephalon, 185, 239
Encystation, of Amœba, 19; of Vorticella, 92, 97
Endoderm, 100, 103
Endogenous cell division, 4
Endoplast, 92
Endopodite, 131, 142, 153
Endosarc, 21
Endoskeleton, 169, 211
Endosperm, 71, 86
Endosporium, 36
Epicoracoid, 222
Epidermis, of Bean, 72, 79, 81; of Fern, 56, 58, 63; of Frog, 261

Epithelium, 251
Epistylis, 97
Eustachian recesses, 166, 196, 248
Evening Primrose, 87
Exoccipital, 217
Exogens, 73
Exopodite, 131, 142, 153
Exoskeleton, of Anodonta, 113, 124; of Crayfish, 127, 140; of Frog, 169; of Lobster, 137, 139
Exosporium, 36
Eye, of Crayfish, 137, 153; of Frog, 194, 244; of Lobster, 137, 153
Eyestalks, 130, 153

F.

FACIALIS, nervus, 190
Femoral artery, 178, 238; vein, 179
Fenestra ovalis, 195, 217, 248
Fermentation, alcoholic, 1, 5, 9, 10; putrefactive, 26
Fertilization, process of, in Anodonta, 112; in Bean, 71; in Chara, 46; in Fern, 61; in Frog, 162; in Hydra, 100
Fibro-vascular bundles, 57, 63, 64, 72, 80
Fibula, 225
Filament, 84
Filum terminale, 192
Fission, 92, 96, 99
Flower, of Bean, 70, 83
Fontanelle, 170
Foot, of Anodonta, 107, 115
Foramen of Munro, 186
Foramen magnum, 216
Fourth ventricle, 185, 240
Fresh water, Crayfish, 127; Laboratory work, 140; Mussel, 107; Laboratory work, 113; Polypes, 98; Laboratory work, 102
Frog, 159; Laboratory work, 196
Fronds, 55, 58, 67
Fungi, 31
Funiculus, 86

M. 18

Cambridge:

PRINTED BY C. J. CLAY, M.A.
AT THE UNIVERSITY PRESS.

A CATALOGUE of EDUCATIONAL BOOKS, Published by MACMILLAN and Co., Bedford Street, Strand, London.

CLASSICAL.

Æschylus.—THE EUMENIDES. The Greek Text, with Introduction, English Notes, and Verse Translation. By BERNARD DRAKE, M.A., late Fellow of King's College, Cambridge. 8vo. 3s. 6d.

Aristotle. — AN INTRODUCTION TO ARISTOTLE'S RHETORIC. With Analysis, Notes, and Appendices. By E. M. COPE, Fellow and Tutor of Trinity Coll. Cambridge. 8vo. 14s.

ARISTOTLE ON FALLACIES ; OR, THE SOPHISTICI ELENCHI. With Translation and Notes by E. POSTE, M.A., Fellow of Oriel College, Oxford. 8vo. 8s. 6d.

Aristophanes.—THE BIRDS. Translated into English Verse, with Introduction, Notes, and Appendices, by B. H. KENNEDY, D.D., Regius Professor of Greek in the University of Cambridge. Crown 8vo. 6s.

Belcher.—SHORT EXERCISES IN LATIN PROSE COMPOSITION AND EXAMINATION PAPERS IN LATIN GRAMMAR, to which is prefixed a Chapter on Analysis of Sentences. By the Rev. H. BELCHER, M.A., Assistant Master in King's College School, London. 18mo. 1s. 6d. Key, 1s. 6d.

Blackie.—GREEK AND ENGLISH DIALOGUES FOR USE IN SCHOOLS AND COLLEGES. By JOHN STUART BLACKIE, Professor of Greek in the Univ. of Edinburgh. Second Edition. Fcap. 8vo. 2s. 6d.

Cicero. — THE SECOND PHILIPPIC ORATION. With Introduction and Notes. From the German of KARL HALM. Edited, with Corrections and Additions, by JOHN E. B. MAYOR. M.A., Fellow and Classical Lecturer of St. John's College, Cambridge. Fourth Edition, revised. Fcap. 8vo. 5s.

THE ORATIONS OF CICERO AGAINST CATILINA. With Notes and an Introduction. From the German of KARL HALM, with additions by A. S. WILKINS, M.A., Owens College, Manchester. New Edition. Fcap. 8vo. 3s. 6d.

5,000, 2, 1876. (Toned)
5,000, 2. 1876. (White).

A

Cicero—*continued.*

THE ACADEMICA OF CICERO. The Text revised and explained by JAMES REID, M.A., Assistant Tutor and late Fellow of Christ's College, Cambridge. Fcap. 8vo. 4*s.* 6*d.*

Demosthenes.—ON THE CROWN, to which is prefixed ÆSCHINÈS AGAINST CTESIPHON. The Greek Text with English Notes. By B. DRAKE, M.A., late Fellow of King's College, Cambridge. Fifth Edition. Fcap. 8vo. 5*s.*

Ellis.—PRACTICAL HINTS ON THE QUANTITATIVE PRONUNCIATION OF LATIN, for the use of Classical Teachers and Linguists. By A. J. ELLIS, B.A., F.R.S. Extra fcap. 8vo. 4*s.* 6*d.*

Goodwin.—SYNTAX OF THE MOODS AND TENSES OF THE GREEK VERB. By W. W. GOODWIN, Ph.D. New Edition, revised. Crown 8vo. 6*s.* 6*d.*

Greenwood.—THE ELEMENTS OF GREEK GRAMMAR, including Accidence, Irregular Verbs, and Principles of Derivation and Composition; adapted to the System of Crude Forms. By J. G. GREENWOOD, Principal of Owens College, Manchester. Fifth Edition. Crown 8vo. 5*s.* 6*d.*

Hodgson.—MYTHOLOGY FOR LATIN VERSIFICATION. A brief Sketch of the Fables of the Ancients, prepared to be rendered into Latin Verse for Schools. By F. HODGSON, B.D., late Provost of Eton. New Edition, revised by F. C. HODGSON, M.A. 18mo. 3*s.*

Homer's Odyssey.—THE NARRATIVE OF ODYSSEUS. With a Commentary by JOHN E. B. MAYOR, M.A., Kennedy Professor of Latin at Cambridge. Part I. Book IX.—XII. Fcap. 8vo. 3*s.*

Horace.—THE WORKS OF HORACE, rendered into English Prose, with Introductions, Running Analysis, and Notes, by JAMES LONSDALE, M.A., and SAMUEL LEE, M.A. Globe 8vo. 3*s.* 6*d.*; gilt edges, 4*s.* 6*d.*

THE ODES OF HORACE IN A METRICAL PARAPHRASE. By R. M. HOVENDEN, B.A., formerly of Trinity College, Cambridge. Extra fcap. 8vo. 4*s.* 6*d.*

Jackson.—FIRST STEPS TO GREEK PROSE COMPOSITION. By BLOMFIELD JACKSON, M.A. Assistant-Master in King's College School, London. 18mo. 1*s.* 6*d.*

Juvenal.—THIRTEEN SATIRES OF JUVENAL. With a Commentary. By JOHN E. B. MAYOR, M.A., Kennedy Professor of Latin at Cambridge. Second Edition, enlarged. Vol. I. Crown 8vo. 7*s.* 6*d.* Or Parts I. and II. Crown 8vo. 3*s.* 6*d.* each.

"*A painstaking and critical edition.*"—SPECTATOR.

Marshall.—A TABLE OF IRREGULAR GREEK VERBS, classified according to the arrangement of Curtius' Greek Grammar. By J. M. MARSHALL, M.A., Fellow and late Lecturer of Brasenose College, Oxford ; one of the Masters in Clifton College. 8vo. cloth. New Edition. 1s.

Mayor (John E. B.)—FIRST GREEK READER. Edited after KARL HALM, with Corrections and large Additions by JOHN E. B. MAYOR, M.A., Fellow and Classical Lecturer of St. John's College, Cambridge. New Edition, revised. Fcap. 8vo. 4s. 6d.

Mayor (John E. B.)—BIBLIOGRAPHICAL CLUE TO LATIN LITERATURE. Edited after HÜBNER, with Large Additions by Professor JOHN E. B. MAYOR. Crown 8vo. 6s. 6d.

"*An extremely useful volume that should be in the hands of all scholars.*"—ATHENÆUM.

Mayor (Joseph B.)—GREEK FOR BEGINNERS. By the Rev. J. B. MAYOR, M.A., Professor of Classical Literature in King's College, London. Part I., with Vocabulary, 1s. 6d. Parts II. and III., with Vocabulary and Index, 3s. 6d., complete in one vol. New Edition. Fcap. 8vo. cloth, 4s. 6d.

Nixon.—PARALLEL EXTRACTS arranged for translation into English and Latin, with Notes on Idioms. By J. E. NIXON, M.A., Classical Lecturer, King's College, London. Part I.—Historical and Epistolary. Crown 8vo. 3s. 6d.

Peile (John, M.A.)—AN INTRODUCTION TO GREEK AND LATIN ETYMOLOGY. By JOHN PEILE, M.A., Fellow and Tutor of Christ's College, Cambridge, formerly Teacher of Sanskrit in the University of Cambridge. Third and Revised Edition. Crown 8vo. 10s. 6d.

"*A very valuable contribution to the science of language.*"—SATURDAY REVIEW.

Plato.—THE REPUBLIC OF PLATO. Translated into English, with an Analysis and Notes, by J. LL. DAVIES, M.A., and D. J. VAUGHAN, M.A. Third Edition, with Vignette Portraits of Plato and Socrates, engraved by JEENS from an Antique Gem. 18mo. 4s. 6d.

Plautus.—THE MOSTELLARIA OF PLAUTUS. With Notes Prolegomena, and Excursus. By WILLIAM RAMSAY, M.A., formerly Professor of Humanity in the University of Glasgow. Edited by Professor GEORGE G. RAMSAY, M.A., of the University of Glasgow. 8vo. 14s.

"*The fruits of that exhaustive research and that ripe and well-digested scholarship which its author brought to bear upon everything that he undertook are visible throughout.*"—PALL MALL GAZETTE.

A 2

Potts, Alex. W., M.A.—HINTS TOWARDS LATIN PROSE COMPOSITION. By Alex. W. Potts, M.A., late Fellow of St. John's College, Cambridge; Assistant Master in Rugby School; and Head Master of the Fettes College, Edinburgh. New Edition, enlarged. Extra fcap. 8vo. cloth. 3s.

Roby.—A GRAMMAR OF THE LATIN LANGUAGE, from Plautus to Suetonius. By H. J. Roby, M.A., late Fellow of St. John's College, Cambridge. In Two Parts. Second Edition. Part I. containing :—Book I. Sounds. Book II. Inflexions. Book III. Word-formation. Appendices. Crown 8vo. 8s. 6d. Part II.—Syntax, Prepositions, &c. Crown 8vo. 10s. 6d.

" *Marked by the clear and practised insight of a master in his art. A book that would do honour to any country.*"—Athenæum.

Rust.—FIRST STEPS TO LATIN PROSE COMPOSITION. By the Rev. G. Rust, M.A. of Pembroke College, Oxford, Master of the Lower School, King's College, London. New Edition. 18mo. 1s. 6d.

Sallust.—CAII SALLUSTII CRISPI CATILINA ET JUGURTHA. For Use in Schools. With copious Notes. By C. Merivale, B.D. New Edition, carefully revised and enlarged. Fcap. 8vo. 4s. 6d. Or separately, 2s. 6d. each.

" *A very good edition, to which the Editor has not only brought scholarship but independent judgment and historical criticism.*"—Spectator.

Tacitus.—THE HISTORY OF TACITUS TRANSLATED INTO ENGLISH. By A. J. Church, M.A., and W. J. Brodribb, M.A. With Notes and a Map. New and Cheaper Edition. Crown 8vo. 6s.

"*A scholarly and faithful translation.*"—Spectator.

TACITUS, THE AGRICOLA AND GERMANIA OF. A Revised Text, English Notes, and Maps. By A. J. Church, M.A., and W. J. Brodribb, M.A. New Edition. Fcap. 8vo. 3s. 6d. Or separately, 2s. each.

" *A model of careful editing, being at once compact, complete, and correct, as well as neatly printed and elegant in style.*"—Athenæum.

TACITUS.—THE ANNALS. Translated, with Notes and Maps, by A. J. Church and W. J. Brodribb. Crown 8vo. 7s. 6d.

THE AGRICOLA AND GERMANIA. Translated into English by A. J. Church, M.A., and W. J. Brodribb, M.A. With Maps and Notes. Extra fcap. 8vo. 2s. 6d.

" *At once readable and exact; may be perused with pleasure by all, and consulted with advantage by the classical student.*"—Athenæum.

Theophrastus. — THE CHARACTERS OF THEO-PHRASTUS. An English Translation from a Revised Text.

With Introduction and Notes. By R. C. JEBB, M.A., Public Orator in the University of Cambridge, and Professor of Greek in the University of Glasgow. Extra fcap. 8vo. 6s. 6d.

"A very handy and scholarly edition of a work which till now has been beset with hindrances and difficulties, but which Mr. Jebb's critical skill and judgment have at length placed within the grasp and comprehension of ordinary readers."—SATURDAY REVIEW.

Thring.—Works by the Rev. E. THRING, M.A., Head Master of Uppingham School.

A LATIN GRADUAL. A First Latin Construing Book for Beginners. New Edition, enlarged, with Coloured Sentence Maps. Fcap. 8vo. 2s. 6d.

A MANUAL OF MOOD CONSTRUCTIONS. Fcap. 8vo. 1s. 6d.

A CONSTRUING BOOK. Fcap. 8vo. 2s. 6d.

Thucydides.—THE SICILIAN EXPEDITION. Being Books VI. and VII. of Thucydides, with Notes. New Edition, revised and enlarged, with Map. By the Rev. PERCIVAL FROST, M.A., late Fellow of St. John's College, Cambridge. Fcap. 8vo. 5s.

"The notes are excellent of their kind. Mr. Frost seldom passes over a difficulty, and what he says is always to the point."—EDUCATIONAL TIMES.

Virgil.—THE WORKS OF VIRGIL RENDERED INTO ENGLISH PROSE, with Notes, Introductions, Running Analysis, and an Index, by JAMES LONSDALE, M.A. and SAMUEL LEE, M.A. Second Edition. Globe 8vo. 3s. 6d.; gilt edges, 4s. 6d.

"A more complete edition of Virgil in English it is scarcely possible to conceive than the scholarly work before us."—GLOBE.

Wright.—Works by J. WRIGHT, M.A., late Head Master of Sutton Coldfield School.

HELLENICA; OR, A HISTORY OF GREECE IN GREEK, as related by Diodorus and Thucydides; being a First Greek Reading Book, with explanatory Notes, Critical and Historical. Third Edition, with a Vocabulary. 12mo. 3s. 6d.

"A good plan well executed."—GUARDIAN.

A HELP TO LATIN GRAMMAR; or, The Form and Use of Words in Latin, with Progressive Exercises. Crown 8vo. 4s. 6d.

THE SEVEN KINGS OF ROME. An Easy Narrative, abridged from the First Book of Livy by the omission of Difficult Passages; being a First Latin Reading Book, with Grammatical Notes. Fifth Edition. Fcap. 8vo. 3s. With Vocabulary, 3s. 6d.

"The Notes are abundant, explicit, and full of such grammatical and other information as boys require."—ATHENÆUM.

FIRST LATIN STEPS; OR, AN INTRODUCTION BY A SERIES OF EXAMPLES TO THE STUDY OF THE LATIN LANGUAGE. Crown 8vo. 5s.

ATTIC PRIMER. Arranged for the Use of Beginners. Extra fcap. 8vo. 4s. 6d.

MATHEMATICS.

Airy.—Works by Sir G. B. AIRY, K.C.B., Astronomer Royal :—
ELEMENTARY TREATISE ON PARTIAL DIFFERENTIAL
EQUATIONS. Designed for the Use of Students in the University. With Diagrams. New Edition. Crown 8vo. cloth.
5s. 6d.

ON THE ALGEBRAICAL AND NUMERICAL THEORY OF
ERRORS OF OBSERVATIONS AND THE COMBINA-
TION OF OBSERVATIONS. New edition, revised. Crown
8vo. cloth. 6s. 6d.

UNDULATORY THEORY OF OPTICS. Designed for the Use of
Students in the University. New Edition. Crown 8vo. cloth.
6s. 6d.

ON SOUND AND ATMOSPHERIC VIBRATIONS. With the
Mathematical Elements of Music. Designed for the Use of Students
of the University. Second Edition, Revised and Enlarged.
Crown 8vo. 9s.

A TREATISE OF MAGNETISM. Designed for the use of
Students in the University. Crown 8vo. 9s. 6d.

Airy (Osmund). — A TREATISE ON GEOMETRICAL
OPTICS. Adapted for the use of the Higher Classes in Schools.
By OSMUND AIRY, B.A., one of the Mathematical Masters in
Wellington College. Extra fcap. 8vo. 3s. 6d.
" *Carefully and lucidly written, and rendered as simple as possible by
the use in all cases of the most elementary form of investigation.*"—
ATHENÆUM.

Bayma.—THE ELEMENTS OF MOLECULAR MECHA-
NICS. By JOSEPH BAYMA, S.J., Professor of Philosophy,
Stonyhurst College. Demy 8vo. cloth. 10s. 6d.

Beasley.—AN ELEMENTARY TREATISE ON PLANE
TRIGONOMETRY. With Examples. By R. D. BEASLEY,
M.A., Head Master of Grantham Grammar School. Fourth
Edition, revised and enlarged. Crown 8vo. cloth. 3s. 6d.

Blackburn (Hugh).— ELEMENTS OF PLAN
TRIGONOMETRY, for the use of the Junior Class of Mathematic
in the University of Glasgow. By HUGH BLACKBURN, M.A.,
Professor of Mathematics in the University of Glasgow. Globe
8vo. 1s. 6d.

Boole.—Works by G. BOOLE, D.C.L., F.R.S., late Professor of
Mathematics in the Queen's University, Ireland.

Boole—*continued*.

A TREATISE ON DIFFERENTIAL EQUATIONS. New and Revised Edition. Edited by I. TODHUNTER. Crown 8vo. cloth. 14*s*.

"*A treatise incomparably superior to any other elementary book on the same subject with which we are acquainted.*"—PHILOSOPHICAL MAGAZINE.

A TREATISE ON DIFFERENTIAL EQUATIONS. Supplementary Volume. Edited by I. TODHUNTER. Crown 8vo. cloth. 8*s*. 6*d*.

This volume contains all that Professor Boole wrote for the purpose of enlarging his treatise on Differential Equations.

THE CALCULUS OF FINITE DIFFERENCES. Crown 8vo. cloth. 10*s*. 6*d*. New Edition, revised by J. F. MOULTON.

"*As an original book by one of the first mathematicians of the age, it is out of all comparison with the mere second-hand compilations which have hitherto been alone accessible to the student.*"—PHILOSOPHICAL MAGAZINE.

Brook - Smith (J.)—ARITHMETIC IN THEORY AND PRACTICE. By J. BROOK-SMITH, M.A., LL.B., St. John's College, Cambridge; Barrister-at-Law; one of the Masters of Cheltenham College. New Edition, revised. Complete, Crown 8vo. 4*s*. 6*d*. Part I. 3*s*. 6*d*.

"*A valuable Manual of Arithmetic of the Scientific kind. The best we have seen.*"—LITERARY CHURCHMAN. "*An essentially practical book, providing very definite help to candidates for almost every kind of competitive examination.*"—BRITISH QUARTERLY.

Cambridge Senate-House Problems and Riders, WITH SOLUTIONS :—

1848-1851.—RIDERS. By JAMESON. 8vo. cloth. 7*s*. 6*d*.
1857.—PROBLEMS AND RIDERS. By CAMPION and WALTON. 8vo. cloth. 8*s*. 6*d*.
1864.—PROBLEMS AND RIDERS. By WALTON and WILKINSON. 8vo. cloth. 10*s*. 6*d*.

CAMBRIDGE COURSE OF ELEMENTARY NATURAL PHILOSOPHY, for the Degree of B.A. Originally compiled by J. C. SNOWBALL, M.A., late Fellow of St. John's College. Fifth Edition, revised and enlarged, and adapted for the Middle-Class Examinations by THOMAS LUND, B.D., Late Fellow and Lecturer of St. John's College, Editor of Wood's Algebra, &c. Crown 8vo. cloth. 5*s*.

Candler.—HELP TO ARITHMETIC. Designed for the use of Schools. By H. CANDLER, M.A., Mathematical Master of Uppingham School. Extra fcap. 8vo. 2*s*. 6*d*.

Cheyne.—Works by C. H. H. CHEYNE, M.A., F.R.A.S.
AN ELEMENTARY TREATISE ON THE PLANETARY
THEORY. With a Collection of Problems. Second Edition.
Crown 8vo. cloth. 6s. 6d.
THE EARTH'S MOTION OF ROTATION. Crown 8vo.
3s. 6d.

Childe.—THE SINGULAR PROPERTIES OF THE ELLIP-
SOID AND ASSOCIATED SURFACES OF THE Nth
DEGREE. By the Rev. G. F. CHILDE, M.A., Author of
" Ray Surfaces," " Related Caustics," &c. 8vo. 10s. 6d.

Christie.—A COLLECTION OF ELEMENTARY TEST-
QUESTIONS IN PURE AND MIXED MATHEMATICS ;
with Answers and Appendices on Synthetic Division, and on the
Solution of Numerical Equations by Horner's Method. By JAMES
R. CHRISTIE, F.R.S., late First Mathematical Master at the
Royal Military Academy, Woolwich. Crown 8vo. cloth. 8s. 6d.

Cuthbertson—EUCLIDIAN GEOMETRY. By FRANCIS
CUTHBERTSON, M.A., LL.D., late Fellow of Corpus Christi
College, Cambridge ; and Head Mathematical Master of the City
of London School. Extra fcap. 8vo. 4s. 6d.

Dalton.—Works by the Rev. T. DALTON, M.A., Assistant
Master of Eton College.

RULES AND EXAMPLES IN ARITHMETIC. New Edition.
18mo. cloth. 2s. 6d. *Answers to the Examples are appended.*

RULES AND EXAMPLES IN ALGEBRA. Part I. 18mo. 2s.
This work is prepared on the same plan as the Arithmetic.

Day.— PROPERTIES OF CONIC SECTIONS PROVED
GEOMETRICALLY. PART I., THE ELLIPSE, with
Problems. By the Rev. H. G. DAY, M.A., Head Master of
Sedbergh Grammar School. Crown 8vo. 3s. 6d.

Dodgson.—AN ELEMENTARY TREATISE ON DETER-
MINANTS, with their Application to Simultaneous Linear
Equations and Algebraical Geometry. By CHARLES L. DODGSON,
M.A., Student and Mathematical Lecturer of Christ Church,
Oxford. Small 4to. cloth. 10s. 6d.
" *A valuable addition to the treatises we possess on Modern Algebra.*"
—EDUCATIONAL TIMES.

Drew.—GEOMETRICAL TREATISE ON CONIC SEC-
TIONS. By W. H. DREW, M.A., St. John's College, Cambridge.
Fifth Edition, enlarged. Crown 8vo. cloth. 5s.
SOLUTIONS TO THE PROBLEMS IN DREW'S CONIC
SECTIONS. Crown 8vo. cloth. 4s. 6d.

Edgar (J. H.) and Pritchard (G. S.)—NOTE-BOOK ON
PRACTICAL SOLID OR DESCRIPTIVE GEOMETRY.
Containing Problems with help for Solutions. By J. H. EDGAR,
M.A., Lecturer on Mechanical Drawing at the Royal School of
Mines, and G. S. PRITCHARD, late Master for Descriptive
Geometry, Royal Military Academy, Woolwich. Third Edition,
revised and enlarged. Globe 8vo. 3s.

Ferrers.—AN ELEMENTARY TREATISE ON TRILINEAR
CO-ORDINATES, the Method of Reciprocal Polars, and the
Theory of Projectors. By the Rev. N. M. FERRERS, M.A., Fellow
and Tutor of Gonville and Caius College, Cambridge. Third
Edition. Crown 8vo. 6s. 6d.

Frost.—Works by PERCIVAL FROST, M.A., formerly Fellow
of St. John's College, Cambridge; Mathematical Lecturer of
King's College.
AN ELEMENTARY TREATISE ON CURVE TRACING. By
PERCIVAL FROST, M.A. 8vo. 12s.
THE FIRST THREE SECTIONS OF NEWTON'S PRINCIPIA.
With Notes and Illustrations. Also a collection of Problems,
principally intended as Examples of Newton's Methods. By
PERCIVAL FROST, M.A. Second Edition. 8vo. cloth. 10s. 6d.

Frost.—SOLID GEOMETRY. By PERCIVAL FROST, M.A.
A New Edition, revised and enlarged, of the Treatise by FROST
and WOLSTENHOLME. In 2 Vols. Vol. I. 8vo. 16s.

Godfray.—Works by HUGH GODFRAY, M.A., Mathematical
Lecturer at Pembroke College, Cambridge.
A TREATISE ON ASTRONOMY, for the Use of Colleges and
Schools. New Edition. 8vo. cloth. 12s. 6d.
AN ELEMENTARY TREATISE ON THE LUNAR THEORY,
with a Brief Sketch of the Problem up to the time of Newton.
Second Edition, revised. Crown 8vo. cloth. 5s. 6d.

Hemming.—AN ELEMENTARY TREATISE ON THE
DIFFERENTIAL AND INTEGRAL CALCULUS, for the
Use of Colleges and Schools. By G. W. HEMMING, M.A.,
Fellow of St. John's College, Cambridge. Second Edition, with
Corrections and Additions. 8vo. cloth. 9s.

Jackson.—GEOMETRICAL CONIC SECTIONS. An Elemen-
tary Treatise in which the Conic Sections are defined as the Plane
Sections of a Cone, and treated by the Method of Projection.
By J. STUART JACKSON, M.A., late Fellow of Gonville and Caius
College, Cambridge. 4s. 6d.

Jellet (John H.)—A TREATISE ON THE THEORY OF
FRICTION. By JOHN H. JELLET, B.D., Senior Fellow of
Trinity College, Dublin; President of the Royal Irish Academy.
8vo. 8s. 6d.

Jones and Cheyne.—ALGEBRAICAL EXERCISES. Progressively arranged. By the Rev. C. A. JONES, M.A., and C. H. CHEYNE, M.A., F.R.A.S., Mathematical Masters of Westminster School. New Edition. 18mo. cloth. 2s. 6d.

Kelland and Tait. — INTRODUCTION TO QUATER-NIONS, with numerous examples. By P. KELLAND, M.A., F.R.S., formerly Fellow of Queen's College, Cambridge; and P. G. TAIT, M.A., formerly Fellow of St. Peter's College, Cambridge; Professors in the department of Mathematics in the University of Edinburgh. Crown 8vo. 7s. 6d.

Kitchener.—A GEOMETRICAL NOTE-BOOK, containing Easy Problems in Geometrical Drawing preparatory to the Study of Geometry. For the Use of Schools. By F. E. KITCHENER, M.A., Mathematical Master at Rugby. New Edition. 4to. 2s.

Morgan.—A COLLECTION OF PROBLEMS AND EXAMPLES IN MATHEMATICS. With Answers. By H. A. MORGAN, M.A., Sadlerian and Mathematical Lecturer of Jesus College, Cambridge. Crown 8vo. cloth. 6s. 6d.

Newton's PRINCIPIA. Edited by Professor Sir W. THOMSON and Professor BLACKBURN. 4to. cloth. 31s. 6d.
  "*Undoubtedly the finest edition of the text of the 'Principia' which has hitherto appeared.*"—EDUCATIONAL TIMES.

Parkinson.—Works by S. PARKINSON, D.D., F.R.S., Tutor and Prælector of St. John's College, Cambridge.
AN ELEMENTARY TREATISE ON MECHANICS. For the Use of the Junior Classes at the University and the Higher Classes in Schools. With a Collection of Examples. Fifth edition, revised. Crown 8vo. cloth. 9s. 6d.
A TREATISE ON OPTICS. Third Edition, revised and enlarged. Crown 8vo. cloth. 10s. 6d.

Phear.—ELEMENTARY HYDROSTATICS. With Numerous Examples. By J. B. PHEAR, M.A., Fellow and late Assistant Tutor of Clare College, Cambridge. Fourth Edition. Crown 8vo. cloth. 5s. 6d.

Pirie.—LESSONS ON RIGID DYNAMICS. By the Rev. G. PIRIE, M.A., Fellow and Tutor of Queen's College, Cambridge. Crown 8vo. 6s.

Pratt.—A TREATISE ON ATTRACTIONS, LAPLACE'S FUNCTIONS, AND THE FIGURE OF THE EARTH. By JOHN H. PRATT, M.A., Archdeacon of Calcutta, Author of "The Mathematical Principles of Mechanical Philosophy." Fourth Edition. Crown 8vo. cloth. 6s. 6d.

Puckle.—AN ELEMENTARY TREATISE ON CONIC SECTIONS AND ALGEBRAIC GEOMETRY. With Numerous Examples and Hints for their Solution; especially designed for the Use of Beginners. By G. H. PUCKLE, M.A. New Edition, revised and enlarged. Crown 8vo. cloth. 7s. 6d.

Rawlinson.—ELEMENTARY STATICS, by the Rev. GEORGE RAWLINSON, M.A. Edited by the Rev. EDWARD STURGES, M.A., of Emmanuel College, Cambridge, and late Professor of the Applied Sciences, Elphinstone College, Bombay. Crown 8vo. cloth. 4*s.* 6*d.*

Reynolds.—MODERN METHODS IN ELEMENTARY GEOMETRY. By E. M. REYNOLDS, M.A., Mathematical Master in Clifton College. Crown 8vo. 3*s.* 6*d.*

Routh.—AN ELEMENTARY TREATISE ON THE DYNAMICS OF THE SYSTEM OF RIGID BODIES. With Numerous Examples. By EDWARD JOHN ROUTH, M.A., late Fellow and Assistant Tutor of St. Peter's College, Cambridge; Examiner in the University of London. Second Edition, enlarged. Crown 8vo. cloth. 14*s.*

WORKS
By the REV. BARNARD SMITH, M.A.,
Rector of Glaston, Rutland, late Fellow and Senior Bursar of St. Peter's College, Cambridge.

ARITHMETIC AND ALGEBRA, in their Principles and Application; with numerous systematically arranged Examples taken from the Cambridge Examination Papers, with especial reference to the Ordinary Examination for the B.A. Degree. Thirteenth Edition, carefully revised. Crown 8vo. cloth. 10*s.* 6*d.*

" *To all those whose minds are sufficiently developed to comprehend the simplest mathematical reasoning, and who have not yet thoroughly mastered the principles of Arithmetic and Algebra, it is calculated to be of great advantage.*"—ATHENÆUM. " *Mr. Smith's work is a most useful publication. The rules are stated with great clearness. The examples are well selected, and worked out with just sufficient detail, without being encumbered by too minute explanations : and there prevails throughout it that just proportion of theory and practice which is the crowning excellence of an elementary work.*" —DEAN PEACOCK.

ARITHMETIC FOR SCHOOLS. New Edition. Crown 8vo. cloth. 4*s.* 6*d.* Adapted from the Author's work on "Arithmetic and Algebra."

" *Admirably adapted for instruction, combining just sufficient theory with a large and well-selected collection of exercises for practice.*"— JOURNAL OF EDUCATION.

A KEY TO THE ARITHMETIC FOR SCHOOLS. Tenth Edition. Crown 8vo. cloth. 8*s.* 6*d.*

EXERCISES IN ARITHMETIC. With Answers. Crown 8vo. limp cloth. 2*s.* 6*d.* Or sold separately, Part I. 1*s.* ; Part II. 1*s.*; Answers, 6*d.*

SCHOOL CLASS-BOOK OF ARITHMETIC. 18mo. cloth. 3*s.* Or sold separately, Parts I. and II. 10*d.* each; Part III. 1*s.*

KEYS TO SCHOOL CLASS-BOOK OF ARITHMETIC, Complete in one volume, 18mo. cloth, 6*s.* 6*d.*; or Parts I., II., and III., 2*s.* 6*d.* each.

Barnard Smith—*continued.*

SHILLING BOOK OF ARITHMETIC FOR NATIONAL AND ELEMENTARY SCHOOLS. 18mo. cloth. Or separately, Part I. 2*d.*; Part II. 3*d.*; Part III. 7*d.* Answers, 6*d*

THE SAME, with Answers complete. 18mo. cloth. 1*s.* 6*d.*

KEY TO SHILLING BOOK OF ARITHMETIC. 18mo. cloth. 4*s.* 6*d.*

EXAMINATION PAPERS IN ARITHMETIC. 18mo. cloth. 1*s.* 6*d.* The same, with Answers, 18mo. 1*s.* 9*d.*

KEY TO EXAMINATION PAPERS IN ARITHMETIC. 18mo. cloth. 4*s.* 6*d.*

THE METRIC SYSTEM OF ARITHMETIC, ITS PRINCIPLES AND APPLICATION, with numerous Examples, written expressly for Standard V. in National Schools. Fourth Edition. 18mo. cloth, sewed. 3*d.*

A CHART OF THE METRIC SYSTEM, on a Sheet, size 42 in. by 34 in. on Roller, mounted and varnished, price 3*s.* 6*d.* Fourth Edition.

" *We do not remember that ever we have seen teaching by a chart more happily carried out.*"—SCHOOL BOARD CHRONICLE.

Also a Small Chart on a Card, price 1*d.*

EASY LESSONS IN ARITHMETIC, combining' Exercises in Reading, Writing, Spelling, and Dictation. Part I. for Standard I. in National Schools. Crown 8vo. 9*d.*

Diagrams for School-room walls in preparation.

" *We should strongly advise everyone to study carefully Mr. Barnard Smith's Lessons in Arithmetic, Writing, and Spelling. A more excellent little work for a first introduction to knowledge cannot well be written. Mr. Smith's larger Text-books on Arithmetic and Algebra are already most favourably known, and he has proved now that the difficulty of writing a text-book which begins* ab ovo *is really surmountable ; but we shall be much mistaken if this little book has not cost its author more thought and mental labour than any of his more elaborate text-books. The plan to combine arithmetical lessons with those in reading and spelling is perfectly novel, and it is worked out in accordance with the aims of our National Schools ; and we are convinced that its general introduction in all elementary schools throughout the country will produce great educational advantages.*"—WESTMINSTER REVIEW.

EXAMINATION CARDS IN ARITHMETIC. (Dedicated to Lord Sandon). With Answers and Hints.

Standards I. and II. in box, 1*s.* 6*d.* Standards III. IV. and V. in boxes, 1*s.* 6*d.* each. Standard VI. in Two Parts, in boxes, 1*s.* 6*d.* each.

A and B papers, of nearly the same difficulty, are given so as to prevent copying, and the Colours of the A and B papers differ in each Standard, and from those of every other Standard, so that a master or mistress can see at a glance whether the children have the proper papers.

Snowball.—THE ELEMENTS OF PLANE AND SPHERI-
CAL TRIGONOMETRY; with the Construction and Use of
Tables of Logarithms. By J. C. SNOWBALL, M.A. Tenth Edition.
Crown 8vo. cloth. 7s. 6d.

SYLLABUS OF PLANE GEOMETRY (corresponding to Euclid,
Books I.—VI.) Prepared by the Association for the Improvement
of Geometrical Teaching. Crown 8vo. 1s.

Tait and Steele.—A TREATISE ON DYNAMICS OF A
PARTICLE. With numerous Examples. By Professor TAIT and
Mr. STEELE. New Edition, enlarged. Crown 8vo. cloth. 10s. 6d.

Tebay.—ELEMENTARY MENSURATION FOR SCHOOLS.
With numerous Examples. By SEPTIMUS TEBAY, B.A., Head
Master of Queen Elizabeth's Grammar School, Rivington. Extra
fcap. 8vo. 3s. 6d.

WORKS

By I. TODHUNTER, M.A., F.R.S.,

Of St. John's College, Cambridge.

"*Mr. Todhunter is chiefly known to students of Mathematics as the
author of a series of admirable mathematical text-books, which possess the
rare qualities of being clear in style and absolutely free from mistakes,
typographical or other.*"—SATURDAY REVIEW.

THE ELEMENTS OF EUCLID. For the Use of Colleges and
Schools. New Edition. 18mo. cloth. 3s. 6d.

MENSURATION FOR BEGINNERS. With numerous Examples.
New Edition. 18mo. cloth. 2s. 6d.

ALGEBRA FOR BEGINNERS. With numerous Examples. New
Edition. 18mo. cloth. 2s. 6d.

KEY TO ALGEBRA FOR BEGINNERS. Crown 8vo. cloth.
6s. 6d.

TRIGONOMETRY FOR BEGINNERS. With numerous Examples.
New Edition. 18mo. cloth. 2s. 6d.

KEY TO TRIGONOMETRY FOR BEGINNERS. Crown 8vo.
8s. 6d.

MECHANICS FOR BEGINNERS. With numerous Examples.
New Edition. 18mo. cloth. 4s. 6d.

ALGEBRA. For the Use of Colleges and Schools. Seventh Edition,
containing two New Chapters and Three Hundred miscellaneous
Examples. Crown 8vo. cloth. 7s. 6d.

KEY TO ALGEBRA FOR THE USE OF COLLEGES AND
SCHOOLS. Crown 8vo. 10s. 6d.

AN ELEMENTARY TREATISE ON THE THEORY OF
EQUATIONS. Third Edition, revised. Crown 8vo. cloth.
7s. 6d.

Todhunter (I.)—*continued.*

PLANE TRIGONOMETRY. For Schools and Colleges. Fifth Edition. Crown 8vo. cloth. 5*s.*

KEY TO PLANE TRIGONOMETRY. Crown 8vo. 10*s.* 6*d.*

A TREATISE ON SPHERICAL TRIGONOMETRY. Third Edition, enlarged. Crown 8vo. cloth. 4*s.* 6*d.*

PLANE CO-ORDINATE GEOMETRY, as applied to the Straight Line and the Conic Sections. With numerous Examples. Fifth Edition, revised and enlarged. Crown 8vo. cloth. 7*s.* 6*d.*

A TREATISE ON THE DIFFERENTIAL CALCULUS. With numerous Examples. Seventh Edition. Crown 8vo. cloth. 10*s.* 6*d.*

A TREATISE ON THE INTEGRAL CALCULUS AND ITS APPLICATIONS. With numerous Examples. Fourth Edition, revised and enlarged. Crown 8vo. cloth. 10*s.* 6*d.*

EXAMPLES OF ANALYTICAL GEOMETRY OF THREE DIMENSIONS. Third Edition, revised. Crown 8vo. cloth. 4*s.*

A TREATISE ON ANALYTICAL STATICS. With numerous Examples. Fourth Edition, revised and enlarged. Crown 8vo. cloth. 10*s.* 6*d.*

A HISTORY OF THE MATHEMATICAL THEORY OF PROBABILITY, from the time of Pascal to that of Laplace. 8vo. 18*s.*

RESEARCHES IN THE CALCULUS OF VARIATIONS, principally on the Theory of Discontinuous Solutions: an Essay to which the Adams Prize was awarded in the University of Cambridge in 1871. 8vo. 6*s.*

A HISTORY OF THE MATHEMATICAL THEORIES OF ATTRACTION, AND THE FIGURE OF THE EARTH, from the time of Newton to that of Laplace. 2 vols. 8vo. 24*s.*

" Such histories are at present more valuable than original work. They at once enable the Mathematician to make himself master of all that has been done on the subject, and also give him a clue to the right method of dealing with the subject in future by showing him the paths by which advance has been made in the past . . . It is with unmingled satisfaction that we see this branch adopted as his special subject by one whose cast of mind and self culture have made him one of the most accurate, as he certainly is the most learned, of Cambridge Mathematicians."—SATURDAY REVIEW.

AN ELEMENTARY TREATISE ON LAPLACE'S, LAMÉ'S, AND BESSEL'S FUNCTIONS. Crown 8vo. 10*s.* 6*d.*

Wilson (J. M.)—ELEMENTARY GEOMETRY. Books I. II. III. Containing the Subjects of Euclid's first Four Books. New Edition, following the Syllabus of the Geometrical Association. By J. M. WILSON, M A., late Fellow of St. John's Col-

Wilson (J. M.)—*continued.*
lege, Cambridge, and Mathematical Master of Rugby School.
Extra fcap. 8vo. 3*s.* 6*d.*

SOLID GEOMETRY AND CONIC SECTIONS. With Appendices on Transversals and Harmonic Division. For the use of Schools. By J. M. WILSON, M.A. Second Edition. Extra fcap. 8vo. 3*s.* 6*d.*

Wilson (W. P.)—A TREATISE ON DYNAMICS. By W. P. WILSON, M.A., Fellow of St. John's College, Cambridge, and Professor of Mathematics in Queen's College, Belfast. 8vo. 9*s.* 6*d.*

"*This treatise supplies a great educational need.*"—EDUCATIONAL TIMES.

Wolstenholme.—A BOOK OF MATHEMATICAL PROBLEMS, on Subjects included in the Cambridge Course. By JOSEPH WOLSTENHOLME, Fellow of Christ's College, sometime Fellow of St. John's College, and lately Lecturer in Mathematics at Christ's College. Crown 8vo. cloth. 8*s.* 6*d.*

"*Judicious, symmetrical, and well arranged.*"— GUARDIAN.

SCIENCE.
ELEMENTARY CLASS-BOOKS.

IT is the intention of the Publishers to produce a complete series of Scientific Manuals, affording full and accurate elementary information, conveyed in clear and lucid English. The authors are well known as among the foremost men of their several departments; and their names form a ready guarantee for the high character of the books. Subjoined is a list of those Manuals that have already appeared, with a short account of each. Others are in active preparation; and the whole will constitute a standard series specially adapted to the requirements of beginners, whether for private study or for school instruction.

ASTRONOMY, by the Astronomer Royal.
POPULAR ASTRONOMY. With Illustrations. By SIR G. B. AIRY, K.C.B., Astronomer Royal. New Edition. 18mo. cloth. 4*s.* 6*d.*

Six lectures, intended "to explain to intelligent persons the principles on which the instruments of an Observatory are constructed; and the

Elementary Class-Books—*continued.*

principles on which the observations made with these instruments are treated for deduction of the distances and weights of the bodies of the Solar System."

ASTRONOMY.

ELEMENTARY LESSONS IN ASTRONOMY. With Coloured Diagram of the Spectra of the Sun, Stars, and Nebulæ, and numerous Illustrations. By J. NORMAN LOCKYER, F.R.S. New Edition. 18mo. 5*s.* 6*d.*

" *Full, clear, sound, and worthy of attention, not only as a popular exposition, but as a scientific ' Index.' "*—ATHENÆUM. " *The most fascinating of elementary books on the Sciences."*—NONCONFORMIST.

QUESTIONS ON LOCKYER'S ELEMENTARY LESSONS IN ASTRONOMY. For the Use of Schools. By JOHN FORBES-ROBERTSON. 18mo. cloth limp. 1*s.* 6*d.*

PHYSIOLOGY.

LESSONS IN ELEMENTARY PHYSIOLOGY. With numerous Illustrations. By T. H. HUXLEY, F.R.S., Professor of Natural History in the Royal School of Mines. New Edition. 18mo. cloth. 4*s.* 6*d.*

" *Pure gold throughout."*—GUARDIAN. " *Unquestionably the clearest and most complete elementary treatise on this subject that we possess in any language."*—WESTMINSTER REVIEW.

QUESTIONS ON HUXLEY'S PHYSIOLOGY FOR SCHOOLS. By T. ALCOCK, M.D. 18mo. 1*s.* 6*d.*

BOTANY.

LESSONS IN ELEMENTARY BOTANY. By D. OLIVER, F.R.S., F.L.S., Professor of Botany in University College, London. With nearly Two Hundred Illustrations. New Edition. 18mo. cloth. 4*s.* 6*d.*

CHEMISTRY.

LESSONS IN ELEMENTARY CHEMISTRY, INORGANIC AND ORGANIC. By HENRY E. ROSCOE, F.R.S., Professor of Chemistry in Owens College, Manchester. With numerous Illustrations and Chromo-Litho of the Solar Spectrum, and of the Alkalies and Alkaline Earths. New Edition. 18mo. cloth. 4*s.* 6*d.*

" *As a standard general text-book it deserves to take a leading place."*—SPECTATOR. " *We unhesitatingly pronounce it the best of all our elementary treatises on Chemistry."*—MEDICAL TIMES.

A SERIES OF CHEMICAL PROBLEMS, prepared with Special Reference to the above, by T. E. THORPE, Ph.D., Professor of Chemistry in the Yorkshire College of Science, Leeds. Adapted for the preparation of Students for the Government, Science, and Society of Arts Examinations. With a Preface by Professor ROSCOE. 18mo. 1*s.* Key. 1*s.*

Elementary Class-Books—*continued.*

POLITICAL ECONOMY.

POLITICAL ECONOMY FOR BEGINNERS. By MILLICENT G. FAWCETT. New Edition. 18mo. 2s. 6d.

"*Clear, compact, and comprehensive.*"—DAILY NEWS. "*The relations of capital and labour have never been more simply or more clearly expounded.*"--CONTEMPORARY REVIEW.

LOGIC.

ELEMENTARY LESSONS IN LOGIC; Deductive and Inductive, with copious Questions and Examples, and a Vocabulary of Logical Terms. By W. STANLEY JEVONS, M.A., Professor of Logic in Owens College, Manchester. New Edition. 18mo. 3s. 6d.

"*Nothing can be better for a school-book.*"—GUARDIAN.

"*A manual alike simple, interesting, and scientific.*"—ATHENÆUM.

PHYSICS.

LESSONS IN ELEMENTARY PHYSICS. By BALFOUR STEWART, F.R.S., Professor of Natural Philosophy in Owens College, Manchester. With numerous Illustrations and Chromoliths of the Spectra of the Sun, Stars, and Nebulæ. New Edition. 18mo. 4s. 6d.

"*The beau-ideal of a scientific text-book, clear, accurate, and thorough.*" EDUCATIONAL TIMES.

PRACTICAL CHEMISTRY.

THE OWENS COLLEGE JUNIOR COURSE OF PRACTICAL CHEMISTRY. By FRANCIS JONES, Chemical Master in the Grammar School, Manchester. With Preface by Professor ROSCOE. With Illustrations. New Edition. 18mo. 2s. 6d.

ANATOMY.

LESSONS IN ELEMENTARY ANATOMY. By ST. GEORGE MIVART, F.R.S., Lecturer in Comparative Anatomy at St. Mary's Hospital. With upwards of 400 Illustrations. 18mo. 6s. 6d.

"*It may be questioned whether any other work on Anatomy contains in like compass so proportionately great a mass of information.*"—LANCET.

"*The work is excellent, and should be in the hands of every student of human anatomy.*"—MEDICAL TIMES.

STEAM.—AN ELEMENTARY TREATISE. By JOHN PERRY, Bachelor of Engineering, Whitworth Scholar, etc., late Lecturer in Physics at Clifton College. With numerous Woodcuts and Numerical Examples and Exercises. 18mo. 4s. 6d.

B

MANUALS FOR STUDENTS.

Flower (W. H.)—AN INTRODUCTION TO THE OSTE-OLOGY OF THE MAMMALIA. Being the substance of the Course of Lectures delivered at the Royal College of Surgeons of England in 1870. By W. H. FLOWER, F.R.S., F.R.C.S., Hunterian Professor of Comparative Anatomy and Physiology, With numerous Illustrations   Globe 8vo. 7s. 6d.

Hooker (Dr.)—THE STUDENT'S FLORA OF THE BRITISH ISLANDS. By J. D. HOOKER, C.B., F.R.S., M.D., D.C.L., President of the Royal Society.   Globe 8vo. 10s. 6d.

"*Cannot fail to perfectly fulfil the purpose for which it is intended.*"— LAND AND WATER.—"*Containing the fullest and most accurate manual of the kind that has yet appeared.*"—PALL MALL GAZETTE.

Oliver (Professor).—FIRST BOOK OF INDIAN BOTANY. By DANIEL OLIVER, F.R.S., F.L.S., Keeper of the Herbarium and Library of the Royal Gardens, Kew, and Professor of Botany in University College, London. With numerous Illustrations. Extra fcap. 8vo. 6s. 6d.

"*It contains a well-digested summary of all essential knowledge pertaining to Indian botany, wrought out in accordance with the best principles of scientific arrangement.*"—ALLEN'S INDIAN MAIL.

Other volumes of these Manuals will follow.

NATURE SERIES.

THE SPECTROSCOPE AND ITS APPLICATIONS. By J. NORMAN LOCKYER, F.R.S. With Coloured Plate and numerous Illustrations. Second Edition. Crown 8vo. 3s. 6d.

THE ORIGIN AND METAMORPHOSES OF INSECTS. By SIR JOHN LUBBOCK, M.P., F.R.S. With numerous Illustrations. Second Edition. Crown 8vo. 3s. 6d.

"*We can most cordially recommend it to young naturalists.*"—ATHE-NÆUM.

THE BIRTH OF CHEMISTRY. By G. F. RODWELL, F.R.A.S., F.C.S., Science Master in Marlborough College. With numerous Illustrations. Crown 8vo. 3s. 6d.

"*We can cordially recommend it to all Students of Chemistry.*"— CHEMICAL NEWS.

THE TRANSIT OF VENUS. By G. FORBES, M.A., Professor of Natural Philosophy in the Andersonian University, Glasgow. Illustrated. Crown 8vo. 3s. 6d.

THE COMMON FROG. By ST. GEORGE MIVART, F.R.S., Lecturer in Comparative Anatomy at St. Mary's Hospital. With numerous Illustrations. Crown 8vo. 3s. 6d.

Nature Series—*continued.*

POLARISATION OF LIGHT. By W. Spottiswoode, F.R.S. With many Illustrations. Crown 8vo. 3s. 6d.

ON BRITISH WILD FLOWERS CONSIDERED IN RELA-TION TO INSECTS. By Sir John Lubbock, Bart., F.R.S. With numerous Illustrations. Second Edition. Crown 8vo. 4s. 6d.

Other volumes to follow.

Ball (R. S., A.M.)—EXPERIMENTAL MECHANICS. A Course of Lectures delivered at the Royal College of Science for Ireland. By R. S. Ball, A.M., Professor of Applied Mathematics and Mechanics in the Royal College of Science for Ireland. Royal 8vo. 16s.

Blanford.—THE RUDIMENTS OF PHYSICAL GEO-GRAPHY FOR THE USE OF INDIAN SCHOOLS; with a Glossary of Technical Terms employed. By H. F. Blanford, F.R.S. Fifth edition, with Illustrations. Globe 8vo. 2s. 6d.

Gordon.—AN ELEMENTARY BOOK ON HEAT. By J. E. H. Gordon, B.A., Gonville and Caius College, Cambridge. Crown 8vo. 2s.

Huxley & Martin.—A COURSE OF PRACTICAL IN-STRUCTION IN ELEMENTARY BIOLOGY. By Professor Huxley, F.R.S., assisted by H. N. Martin, M.B., D.Sc. Crown 8vo. 6s.

SCIENCE PRIMERS FOR ELEMENTARY SCHOOLS.

In these Primers the authors have aimed, not so much to give informa-tion, as to endeavour to discipline the mind in a way which has not hitherto been customary, by bringing it into immediate contact with Nature herself. For this purpose a series of simple experiments (to be performed by the teacher) has been devised, leading up to the chief truths of each Science. Thus the power of observation in the pupils will be awakened and strengthened. Each Manual is copiously illustrated, and appended are lists of all the necessary apparatus, with prices, and directions as to how they may be obtained. Professor Huxley's introduc-tory volume has been delayed through the illness of the author, but it is now expected to appear very shortly. " They are wonderfully clear and lucid in their instruction, simple in style, and admirable in plan."—
Educational Times.

PRIMER OF CHEMISTRY. By H. E. Roscoe, Professor of Chemistry in Owens College, Manchester. With numerous Illus-trations. 18mo. 1s. New Edition.

PRIMER OF PHYSICS. By Balfour Stewart, Professor of Natural Philosophy in Owens College, Manchester. With numerous Illustrations. 18mo. 1s. New Edition.

EDUCATIONAL BOOKS.

PRIMER OF PHYSICAL GEOGRAPHY. By ARCHIBALD
GEIKIE, F.R.S., Murchison-Professor of Geology and Mineralogy
at Edinburgh. With numerous Illustrations. New Edition.
18mo. 1s.

PRIMER OF GEOLOGY. By PROFESSOR GEIKIE, F.R.S. With
numerous Illustrations. New Edition. 18mo. cloth. 1s.

PRIMER OF PHYSIOLOGY. By MICHAEL FOSTER, M.D.,
F.R.S. With numerous Illustrations. New Edition. 18mo. 1s.

PRIMER OF ASTRONOMY. By J. NORMAN LOCKYER, F.R.S.
With numerous Illustrations. New Edition. 18mo. 1s.

PRIMER OF BOTANY. By J. D. HOOKER, C.B. F.R.S., President of the Royal Society. With numerous Illustrations. 18mo.
1s.

In preparation:—

INTRODUCTORY. By PROFESSOR HUXLEY. &r. &c.

MISCELLANEOUS.

Abbott.—A SHAKESPEARIAN GRAMMAR. An Attempt to
illustrate some of the Differences between Elizabethan and Modern
English. By the Rev. E. A. ABBOTT, M.A., Head Master of the
City of London School. For the Use of Schools. New and
Enlarged Edition. Extra fcap. 8vo. 6s.

"*A critical inquiry, conducted with great skill and knowledge, and
with all the appliances of modern philology....*"—PALL MALL
GAZETTE. "*Valuable not only as an aid to the critical study of
Shakespeare, but as tending to familiarize the reader with Elizabethan
English in general.*"—ATHENÆUM.

Baldwin.—INTRODUCTION TO PRACTICAL FARMING
FOR THE USE OF SCHOOLS. By T. BALDWIN, M.R.I.A.
Superintendent of the Agricultural Department of National Educa-
tion in Ireland. 18mo. 1s. 6d.

Barker.—FIRST LESSONS IN THE PRINCIPLES OF
COOKING. By LADY BARKER. 18mo. 1s.

"*An unpretending but invaluable little work.... The plan is
admirable in its completeness and simplicity; it is hardly possible that
anyone who can read at all can fail to understand the practical lessons on
bread and beef, fish and vegetables; while the explanation of the chemical
composition of our food must be intelligible to all who possess sufficient
education to follow the argument, in which the fewest possible technical
terms are used.*"—SPECTATOR.

Berners.—FIRST LESSONS ON HEALTH. By J. Ber-
NERS. 18mo. 1s. Fourth Edition.

Besant.—STUDIES IN EARLY FRENCH POETRY. By
WALTER BESANT, M.A. Crown 8vo. 8s. 6d.
" *In one moderately sized volume he has contrived to introduce us to the*
very best, if not to all of the early French poets."—ATHENÆUM.

Breymann.—Works by HERMANN BREYMANN, Ph.D., late
Lecturer on French Language and Literature at Owens College,
Manchester, and now Professor of Philology in the University of
Munich.

A FRENCH GRAMMAR BASED ON PHILOLOGICAL
PRINCIPLES. Second Edition. Extra fcap. 8vo. 4s. 6d.
" *We dismiss the work with every expression of satisfaction. It can-*
not fail to be taken into use by all schools which endeavour to make the
study of French a means towards the higher culture."—EDUCATIONAL
TIMES. "*A good, sound, valuable philological grammar. The author*
presents the pupil by his method and by detail, with an enormous amount
of information about French not usually to be found in grammars, and
the information is all of it of real practical value to the student who
really wants to know French well, and to understand its spirit."—
SCHOOL BOARD CHRONICLE.

FIRST FRENCH EXERCISE BOOK. Extra fcap. 8vo. 4s. 6d.

SECOND FRENCH EXERCISE BOOK. Extra fcap. 8vo. 2s. 6d.

Calderwood.—HANDBOOK OF MORAL PHILOSOPHY.
By the Rev. HENRY CALDERWOOD, LL.D., Professor of Moral
Philosophy, University of Edinburgh. Fourth Edition. Crown
8vo. 6s.
" *A compact and useful work . . . . will be an assistance to many*
students outside the author's own University."—GUARDIAN.

Delamotte.—A BEGINNER'S DRAWING BOOK. By P. H.
DELAMOTTE, F.S.A. Progressively arranged. New Edition,
improved. Crown 8vo. 3s. 6d.
" *A concise, simple, and thoroughly practical work.*"—GUARDIAN.

Fawcett.—TALES IN POLITICAL ECONOMY. By MILLI-
CENT GARRETT FAWCETT. Globe 8vo, 3s.
" *The idea is a good one, and it is quite wonderful what a mass of*
economic teaching the author manages to compress into a small space."—
ATHENÆUM.

Goldsmith.—THE TRAVELLER, or a Prospect of Society;
and THE DESERTED VILLAGE. By OLIVER GOLDSMITH.
With Notes Philological and Explanatory, by J. W. HALES, M.A.
Crown 8vo. 6d.

Hales.—LONGER ENGLISH POEMS, with Notes, Philological and Explanatory, and an Introduction on the Teaching of English. Chiefly for use in Schools. Edited by J. W. HALES, M.A., Lecturer in English Literature and Classical Composition at King's College School, London, &c. &c. Third Edition. Extra fcap. 8vo. 4s. 6d.

" *The notes are very full and good, and the book, edited by one of our most cultivated English scholars, is probably the best volume of selections ever made for the use of English schools.*"—PROFESSOR MORLEY'S *First Sketch of English Literature.*

Helfenstein (James).—A COMPARATIVE GRAMMAR OF THE TEUTONIC LANGUAGES. By JAMES HELFEN-STEIN, Ph.D. 8vo. 18s.

Hole.—A GENEALOGICAL STEMMA OF THE KINGS OF ENGLAND AND FRANCE. By the Rev. C. HOLE. On Sheet. 1s.

Jephson.—SHAKESPEARE'S "TEMPEST." With Glossarial and Explanatory Notes. By the Rev. J. M. JEPHSON. Second Edition. 18mo. 1s.

Literature Primers.—Edited by JOHN RICHARD GREEN. Author of "A Short History of the English People."

ENGLISH GRAMMAR. By the Rev. R. MORRIS, LL.D., President of the Philological Society. 18mo. cloth. 1s.

" *A work quite precious in its way. . . . An excellent English Grammar for the lowest form.*"—EDUCATIONAL TIMES.

THE CHILDREN'S TREASURY OF ENGLISH SONG. Selected and arranged with Notes by FRANCIS TURNER PALGRAVE. In Two Parts. 18mo. 1s. each.

ENGLISH LITERATURE. By the Rev. STOPFORD BROOKE, M.A. 18mo. 1s.

In preparation :—
> LATIN LITERATURE. By the Rev. Dr. FARRAR, F.R.S.
> GREEK LITERATURE. By PROFESSOR JEBB, M.A.
> SHAKSPERE. By PROFESSOR DOWDEN.
> PHILOLOGY. By J. PEILE, M.A.
> BIBLE PRIMER. By G. GROVE, D.C.L.
> CHAUCER. By F. J. FURNIVALL, M.A.
> GREEK ANTIQUITIES. By the Rev. J. P. MAHAFFY, M.A.

Martin.—THE POET'S HOUR : Poetry Selected and Arranged for Children. By FRANCES MARTIN. Second Edition. 18mo. 2s. 6d.

SPRING-TIME WITH THE POETS. Poetry selected by FRANCES MARTIN. Second Edition. 18mo. 3s. 6d.

Masson (Gustave).—A COMPENDIOUS DICTIONARY OF THE FRENCH LANGUAGE (French-English and English-French). Followed by a List of the Principal Diverging Derivations, and preceded by Chronological and Historical Tables. By GUSTAVE MASSON, Assistant-Master and Librarian, Harrow School. Second Edition. Square half-bound, 6s.

"*By many degrees the most useful Dictionary that the student can obtain.*"—EDUCATIONAL TIMES.

"*A book which any student, whatever may be the degree of his advancement in the language, would do well to have on the table close at hand while he is reading.*"—SATURDAY REVIEW.

Morris.—Works by the Rev. R. MORRIS, LL.D., Lecturer on English Language and Literature in King's College School.

HISTORICAL OUTLINES OF ENGLISH ACCIDENCE, comprising Chapters on the History and Development of the Language, and on Word-formation. Third Edition. Extra fcap. 8vo. 6s.

"*It makes an era in the study of the English tongue.*"—SATURDAY REVIEW. "*A genuine and sound book.*"—ATHENÆUM.

ELEMENTARY LESSONS IN HISTORICAL ENGLISH GRAMMAR, Containing Accidence and Word-formation. Second Edition. 18mo. 2s. 6d.

PRIMER OF ENGLISH GRAMMAR. 18mo. 1s.

Oliphant.—THE SOURCES OF STANDARD ENGLISH. By J. KINGTON OLIPHANT. Extra fcap. 8vo. 6s.

"*Mr. Oliphant's book is, to our mind, one of the ablest and most scholarly contributions to our standard English we have seen for many years. . . . The arrangement of the work and its indices make it invaluable as a work of reference, and easy alike to study and to store, when studied, in the memory.*"—SCHOOL BOARD CHRONICLE. "*Comes nearer to a history of the English language than anything that we have seen since such a history could be written without confusion and contradictions.*"—SATURDAY REVIEW.

Oppen.—FRENCH READER. For the Use of Colleges and Schools. Containing a graduated Selection from modern Authors in Prose and Verse; and copious Notes, chiefly Etymological. By EDWARD A. OPPEN. Fcap. 8vo. cloth. 4s. 6d.

Otté.—SCANDINAVIAN HISTORY. By E. C. OTTÉ. With Maps. Globe 8vo. 6s.

"*A readable, well-arranged, complete, and accurate volume.*"—LITERARY REVIEW.

Palgrave.—THE CHILDREN'S TREASURY OF ENGLISH SONG. Selected and Arranged with Notes by FRANCIS TURNER PALGRAVE. In Two Parts. 18mo. 1s. each.

"*While indeed a treasure for intelligent children, it is also a work which many older folk will be glad to have.*"—SATURDAY REVIEW.

Pylodet.—NEW GUIDE TO GERMAN CONVERSATION: containing an Alphabetical List of nearly 800 Familiar Words followed by Exercises, Vocabulary of Words in frequent use, Familiar Phrases and Dialogues; a Sketch of German Literature, Idiomatic Expressions, &c. By L. PYLODET. 18mo. cloth limp. 2s. 6d.

Reading Books.—Adapted to the English and Scotch Codes for 1875. Bound in Cloth.

PRIMER. 18mo. (48 pp.) 2*d*.

BOOK	I. for Standard	I.	18mo.	(96 pp.)	3*d*.
	II.	II.	18mo.	(144 pp.)	4*d*.
	III.	III.	18mo.	(160 pp.)	6*d*.
	IV.	IV.	18mo.	(176 pp.)	8*d*.
	V.	V.	18mo.	(380 pp.)	1*s*.
	VI.	VI.	Crown 8vo.	(430 pp.)	2*s*.

Book VI. is fitted for higher Classes, and as an Introduction to English Literature.

Sonnenschein and Meiklejohn.—THE ENGLISH METHOD OF TEACHING TO READ. By A. SONNENSCHEIN and J. M. D. MEIKLEJOHN, M.A. Fcap. 8vo.

COMPRISING :

THE NURSERY BOOK, containing all the Two-Letter Words in the Language. 1*d*. (Also in Large Type on Sheets for School Walls. 5*s*.)

THE FIRST COURSE, consisting of Short Vowels with Single Consonants. 3*d*.

THE SECOND COURSE, with Combinations and Bridges, consisting of Short Vowels with Double Consonants. 4*d*.

THE THIRD AND FOURTH COURSES, consisting of Long Vowels, and all the Double Vowels in the Language. 6*d*.

" These are admirable books, because they are constructed on a principle, and that the simplest principle on which it is possible to learn to read English."—SPECTATOR.

Taylor.—WORDS AND PLACES ; or, Etymological Illustrations of History, Ethnology, and Geography. By the Rev. ISAAC TAYLOR, M.A. Third and cheaper Edition, revised and compressed. With Maps. Globe 8vo. 6*s*.

Already been adopted by many teachers, and prescribed as a text-book in the Cambridge Higher Examinations for Women.

Thring.—Works by EDWARD THRING, M.A., Head Master of Uppingham.

THE ELEMENTS OF GRAMMAR TAUGHT IN ENGLISH, with Questions. Fourth Edition. 18mo. 2*s*.

THE CHILD'S GRAMMAR. Being the Substance of "The Elements of Grammar taught in English," adapted for the Use of Junior Classes. A New Edition. 18mo. 1*s*.

SCHOOL SONGS. A Collection of Songs for Schools. With the Music arranged for four Voices. Edited by the Rev. E. THRING and H. RICCIUS. Folio. 7*s*. 6*d*.

Trench (Archbishop).—Works by R. C. TRENCH, D.D., Archbishop of Dublin.

HOUSEHOLD BOOK OF ENGLISH POETRY. Selected and Arranged, with Notes. Extra fcap. 8vo. 5s. 6d. Second Edition. *" The Archbishop has conferred in this delightful volume an important gift on the whole English-speaking population of the world."*—PALL MALL GAZETTE.

ON THE STUDY OF WORDS. Lectures addressed (originally) to the Pupils at the Diocesan Training School, Winchester. Fifteenth Edition, revised. Fcap. 8vo. 4s. 6d.

ENGLISH, PAST AND PRESENT. Ninth Edition, revised and improved. Fcap. 8vo. 5s.

A SELECT GLOSSARY OF ENGLISH WORDS, used formerly in Senses Different from their Present. Fourth Edition, enlarged. Fcap. 8vo. 4s. 6d.

Vaughan (C. M.)—A SHILLING BOOK OF WORDS FROM THE POETS. By C. M. VAUGHAN. 18mo. cloth.

Whitney.—Works by WILLIAM D. WHITNEY, Professor of Sanskrit and Instructor in Modern Languages in Yale College; first President of the American Philological Association, and hon. member of the Royal Asiatic Society of Great Britain and Ireland; and Correspondent of the Berlin Academy of Sciences.

A COMPENDIOUS GERMAN GRAMMAR. Crown 8vo. 6s.

A GERMAN READER IN PROSE AND VERSE, with Notes and Vocabulary. Crown 8vo. 7s. 6d.

Yonge (Charlotte M.)—THE ABRIDGED BOOK OF GOLDEN DEEDS. A Reading Book for Schools and General Readers. By the Author of "The Heir of Redclyffe." 18mo. cloth. 1s.

HISTORY.

Freeman (Edward A.)—OLD-ENGLISH HISTORY. By EDWARD A. FREEMAN, D.C.L., late Fellow of Trinity College, Oxford. With Five Coloured Maps. Fourth Edition. Extra fcap. 8vo. half-bound. 6s.

"I have, I hope," the author says, "shown that it is perfectly easy to teach children, from the very first, to distinguish true history alike from legend and from wilful invention, and also to understand the nature of historical authorities and to weigh one statement against another. I have throughout striven to connect the history of England with the general history of civilised Europe, and I have especially tried to make the book serve as an incentive to a more accurate study of historical geography." In the present edition the whole has been carefully revised,

and such improvements as suggested themselves have been introduced.
" *The book indeed is full of instruction and interest to students of all ages, and he must be a well-informed man indeed who will not rise from its perusal with clearer and more accurate ideas of a too much neglected portion of English History.*"—SPECTATOR.

Green.—A SHORT HISTORY OF THE ENGLISH PEOPLE.
By JOHN RICHARD GREEN. With Coloured Maps, Genealogical Tables, and Chronological Annals. Crown 8vo. 8s. 6d. Thirty-fourth Thousand.

" *Stands alone as the one general history of the country, for the sake of which all others, if young and old are wise, will be speedily and surely set aside.*"—ACADEMY.

Historical Course for Schools.—Edited by EDWARD A. FREEMAN, D.C.L., late Fellow of Trinity College, Oxford.
The object of the present series is to put forth clear and correct views of history in simple language, and in the smallest space and cheapest form in which it could be done. It is meant in the first place for Schools ; but it is often found that a book for schools proves useful for other readers as well, and it is hoped that this may be the case with the little books the first instalment of which is now given to the world.

I. GENERAL SKETCH OF EUROPEAN HISTORY. By EDWARD A. FREEMAN, D.C.L. Fourth Edition. 18mo. cloth. 3s. 6d.
" *It supplies the great want of a good foundation for historical teaching. The scheme is an excellent one, and this instalment has been executed in a way that promises much for the volumes that are yet to appear.*"—EDUCATIONAL TIMES.

II. HISTORY OF ENGLAND. By EDITH THOMPSON. Fifth Edition. 18mo. 2s. 6d.
" *Freedom from prejudice, simplicity of style, and accuracy of statement, are the characteristics of this little volume. It is a trustworthy text-book and likely to be generally serviceable in schools.*"—PALL MALL GAZETTE.
" *Upon the whole, this manual is the best sketch of English history for the use of young people we have yet met with.*"—ATHENÆUM.

III. HISTORY OF SCOTLAND. By MARGARET MACARTHUR. 18mo. 2s.
" *An excellent summary, unimpeachable as to facts, and putting them in the clearest and most impartial light attainable.*"—GUARDIAN. " *Miss Macarthur has performed her task with admirable care, clearness, and fulness, and we have now for the first time a really good School History of Scotland.*"—EDUCATIONAL TIMES.

IV. HISTORY OF ITALY. By the Rev. W. HUNT, M.A. 18mo. 3s.
" *It possesses the same solid merit as its predecessors . . . . the same scrupulous care about fidelity in details. . . . It is distinguished, too, by*

HISTORY.

Historical Course for Schools—*continued.*

information on art, architecture, and social politics, in which the writer's grasp is seen by the firmness and clearness of his touch."—EDUCATIONAL TIMES.

V. HISTORY OF GERMANY. By J. SIME, M.A. 18mo. 3s.
 " A remarkably clear and impressive History of Germany. Its great events are wisely kept as central figures, and the smaller events are carefully kept, not only subordinate and subservient, but most skilfully woven into the texture of the historical tapestry presented to the eye."—STANDARD.

VI. HISTORY OF AMERICA. By JOHN A. DOYLE. With Maps. 18mo. 4s. 6d.
 " Mr. Doyle has performed his task with admirable care, fulness, and clearness, and for the first time we have for schools an accurate and interesting history of America, from the earliest to the present time."—STANDARD.

The following will shortly be issued:—
 FRANCE. By CHARLOTTE M. YONGE.
 GREECE. By J. ANNAN BRYCE, B.A.

History Primers.—Edited by JOHN RICHARD GREEN. Author of "A Short History of the English People."

ROME. By the Rev. M. Creighton, M.A., Fellow and Tutor of Merton College, Oxford. With Eleven Maps. 18mo. 1s.
 " The Author has been curiously successful in telling in an intelligent way the story of Rome from first to last."—SCHOOL BOARD CHRONICLE.

GREECE. By C. A. Fyffe, M.A., Fellow and late Tutor of University College, Oxford. With Five Maps. 18mo. 1s.
 " We give our unqualified praise to this little manual."—SCHOOL-MASTER.

In preparation:—
 EUROPE. By E. A. FREEMAN, D.C.L., LL.D.
 ENGLAND. By J. R. GREEN, M.A.
 FRANCE. By CHARLOTTE M. YONGE.
 GEOGRAPHY. By GEORGE GROVE, D.C.L.

Michelet.—A SUMMARY OF MODERN HISTORY. Translated from the French of M. Michelet, and continued to the Present Time, by M. C. M. Simpson. Globe 8vo. 4s. 6d.

 " We are glad to see one of the ablest and most useful summaries of European history put into the hands of English readers. The translation is excellent."—STANDARD.

Yonge (Charlotte M.)—A PARALLEL HISTORY OF FRANCE AND ENGLAND: consisting of Outlines and Dates. By CHARLOTTE M. YONGE, Author of "The Heir of Redclyffe, "Cameos of English History," &c. &c. Oblong 4to. 3s. 6d.

 " We can imagine few more really advantageous courses of historical study for a young mind than going carefully and steadily through Miss Yonge's excellent little book."—EDUCATIONAL TIMES.

Yonge (Charlotte M.)—*continued.*

CAMEOS FROM ENGLISH HISTORY. From Rollo to Edward II. By the Author of "The Heir of Redclyffe." Extra fcap. 8vo. Third Edition, enlarged. 5*s.*

A book for young people just beyond the elementary histories of England, and able to enter in some degree into the real spirit of events, and to be struck with characters and scenes presented in some relief. "Instead of dry details, we have living pictures, faithful, vivid, and striking."—NONCONFORMIST.

A SECOND SERIES OF CAMEOS FROM ENGLISH HISTORY. THE WARS IN FRANCE. Third Edition. Extra fcap. 8vo. 5*s.*

"Though mainly intended for young readers, they will, if we mistake not, be found very acceptable to those of more mature years, and the life and reality imparted to the dry bones of history cannot fail to be attractive to readers of every age."—JOHN BULL.

EUROPEAN HISTORY. Narrated in a Series of Historical Selections from the Best Authorities. Edited and arranged by E. M. SEWELL and C. M. YONGE. First Series, 1003—1154. Third Edition. Crown 8vo. 6*s.* Second Series, 1088—1228. Crown 8vo. 6*s.* Third Edition.

"We know of scarcely anything which is so likely to raise to a higher level the average standard of English education."—GUARDIAN.

DIVINITY.

** For other Works by these Authors, see THEOLOGICAL CATALOGUE.

Abbott (Rev. E. A.)—BIBLE LESSONS. By the Rev. E. A. ABBOTT, M.A., Head Master of the City of London School. Second Edition. Crown 8vo. 4*s.* 6*d.*

"Wise, suggestive, and really profound initiation into religious thought."—GUARDIAN. *"I think nobody could read them without being both the better for them himself, and being also able to see how this difficult duty of imparting a sound religious education may be effected."*—BISHOP OF ST. DAVID'S AT ABERGWILLY.

Arnold.— A BIBLE-READING FOR SCHOOLS. The GREAT PROPHECY OF ISRAEL'S RESTORATION (Isaiah, Chapters 40—66). Arranged and Edited for Young Learners. By MATTHEW ARNOLD, D.C.L., formerly Professor of Poetry in the University of Oxford, and Fellow of Oriel. Fourth Edition. 18mo. cloth. 1*s.*

"There can be no doubt that it will be found excellently calculated to further instruction in Biblical literature in any school into which it may be introduced; and we can safely say that whatever school uses the book, it will enable its pupils to understand Isaiah, a great advantage compared with other establishments which do not avail themselves of it."—TIMES.

Arnold.—ISAIAH XL.—LXVI. With the Shorter Prophecies allied to it. Arranged and Edited with Notes by MATTHEW ARNOLD. Crown 8vo. 5s.

Golden Treasury Psalter.—Students' Edition. Being an Edition of "The Psalms Chronologically Arranged, by Four Friends," with briefer Notes. 18mo. 3s. 6d.

Hardwick.—A HISTORY OF THE CHRISTIAN CHURCH. Middle Age. From Gregory the Great to the Excommunication of Luther. Edited by WILLIAM STUBBS, M.A., Regius Professor of Modern History in the University of Oxford. With Four Maps constructed for this work by A. KEITH JOHNSTON. Fourth Edition. Crown 8vo. 10s. 6d.

For this edition Professor Stubbs has carefully revised both text and notes, making such corrections of facts, dates, and the like as the results of recent research warrant. The doctrinal, historical, and generally speculative views of the late author have been preserved intact. "As a manual for the student of ecclesiastical history in the Middle Ages, we know no English work which can be compared to Mr. Hardwick's book."—GUARDIAN.

A HISTORY OF THE CHRISTIAN CHURCH DURING THE REFORMATION. By ARCHDEACON HARDWICK. Fourth Edition. Edited by Professor STUBBS. Crown 8vo. 10s. 6d.

Maclear.—Works by the Rev. G. F. MACLEAR, D.D., Head Master of King's College School.

A CLASS-BOOK OF OLD TESTAMENT HISTORY. Eighth Edition, with Four Maps. 18mo. cloth. 4s. 6d.

"A careful and elaborate though brief compendium of all that modern research has done for the illustration of the Old Testament. We know of no work which contains so much important information in so small a compass."—BRITISH QUARTERLY REVIEW.

A CLASS-BOOK OF NEW TESTAMENT HISTORY, including the Connexion of the Old and New Testament. With Four Maps. Fifth Edition. 18mo. cloth. 5s. 6d.

"A singularly clear and orderly arrangement of the Sacred Story. His work is solidly and completely done."—ATHENÆUM.

A SHILLING BOOK OF OLD TESTAMENT HISTORY, for National and Elementary Schools. With Map. 18mo. cloth. New Edition.

A SHILLING BOOK OF NEW TESTAMENT HISTORY, for National and Elementary Schools. With Map. 18mo. cloth. New Edition.

These works have been carefully abridged from the author's larger manuals.

Maclear—*continued.*

CLASS-BOOK OF THE CATECHISM OF THE CHURCH OF
ENGLAND. New and Cheaper Edition. 18mo. cloth. 1*s.* 6*d.*
"*It is indeed the work of a scholar and divine, and as such, though
extremely simple, it is also extremely instructive. There are few clergymen who would not find it useful in preparing candidates for Confirmation; and there are not a few who would find it useful to themselves
as well.*"—LITERARY CHURCHMAN.

A FIRST CLASS-BOOK OF THE CATECHISM OF THE
CHURCH OF ENGLAND, with Scripture Proofs, for Junior
Classes and Schools. 18mo. 6*d.* New Edition.

A MANUAL OF INSTRUCTION FOR CONFIRMATION AND
FIRST COMMUNION. With Prayers and Devotions. Royal
32mo. cloth extra, red edges. 2*s.*
"*It is earnest, orthodox, and affectionate in tone. The form of self-
examination is particularly good.*"—JOHN BULL.

THE ORDER OF CONFIRMATION, WITH PRAYERS AND
DEVOTIONS. 32mo. 6*d.*

FIRST COMMUNION, WITH PRAYERS AND DEVOTIONS
FOR THE NEWLY CONFIRMED. 32mo. 6*d.*

Maurice.—THE LORD'S PRAYER, THE CREED, AND
THE COMMANDMENTS. A Manual for Parents and School-
masters. To which is added the Order of the Scriptures. By the
Rev. F. DENISON MAURICE, M.A. 18mo. cloth limp. 1*s.*

Procter.—A HISTORY OF THE BOOK OF COMMON
PRAYER, with a Rationale of its Offices. By FRANCIS PROCTER,
M.A. Twelfth Edition, revised and enlarged. Crown 8vo.
10*s.* 6*d.*

Procter and Maclear.—AN ELEMENTARY INTRO-
DUCTION TO THE BOOK OF COMMON PRAYER.
Re-arranged and supplemented by an Explanation of the Morning
and Evening Prayer and the Litany. By the Rev. F. PROCTER
and the Rev. G. F. MACLEAR. New Edition. 18mo. 2*s.* 6*d.*

**Psalms of David Chronologically Arranged. By
Four Friends.** An Amended Version, with Historical
Introduction and Explanatory Notes. Second and Cheaper
Edition, with Additions and Corrections. Crown 8vo. 8*s.* 6*d.*
"*One of the most instructive and valuable books that has been published
for many years.*"—SPECTATOR.

Ramsay.—THE CATECHISER'S MANUAL; or, the Church
Catechism Illustrated and Explained, for the use of Clergymen,
Schoolmasters, and Teachers. By the Rev. ARTHUR RAMSAY,
M.A. Second Edition. 18mo. 1*s.* 6*d.*

Simpson.—AN EPITOME OF THE HISTORY OF THE CHRISTIAN CHURCH. By WILLIAM SIMPSON, M.A. Fifth Edition. Fcap. 8vo. 3*s.* 6*d.*

Swainson.—A HANDBOOK to BUTLER'S ANALOGY. By C. A. SWAINSON, D.D., Canon of Chichester. Crown 8vo. 1*s.* 6*d.*

Trench.—SYNONYMS OF THE NEW TESTAMENT. By R. CHENEVIX TRENCH, D.D., Archbishop of Dublin. New Edition, enlarged. 8vo. cloth. 12*s.*

Westcott.—Works by BROOKE FOSS WESTCOTT, B.D., Canon of Peterborough.

A GENERAL SURVEY OF THE HISTORY OF THE CANON OF THE NEW TESTAMENT DURING THE FIRST FOUR CENTURIES. Fourth Edition. With Preface on "Supernatural Religion." Crown 8vo. 10*s.* 6*d.*

" *Theological students, and not they only, but the general public, owe a deep debt of gratitude to Mr. Westcott for bringing this subject fairly before them in this candid and comprehensive essay. . . . . As a theological work it is at once perfectly fair and impartial, and imbued with a thoroughly religious spirit; and as a manual it exhibits, in a lucid form and in a narrow compass, the results of extensive research and accurate thought. We cordially recommend it.*"—SATURDAY REVIEW.

INTRODUCTION TO THE STUDY OF THE FOUR GOSPELS. Fifth Edition. Crown 8vo. 10*s.* 6*d.*

" *To learning and accuracy which commands respect and confidence, he unites what are not always to be found in union with these qualities, the no less valuable faculties of lucid arrangement and graceful and facile expression.*"—LONDON QUARTERLY REVIEW.

THE BIBLE IN THE CHURCH. A Popular Account of the Collection and Reception of the Holy Scriptures in the Christian Churches. New Edition. 18mo. cloth. 4*s.* 6*d.*

"*We would recommend every one who loves and studies the Bible to read and ponder this exquisite little book. Mr. Westcott's account of the 'Canon' is* true history *in its highest sense.*"—LITERARY CHURCHMAN.

THE GOSPEL OF THE RESURRECTION. Thoughts on its Relation to Reason and History. New Edition. Crown 8vo. 6*s.*

Wilson.—THE BIBLE STUDENT'S GUIDE to the more Correct Understanding of the English translation of the Old Testament, by reference to the Original Hebrew. By WILLIAM WILSON, D.D., Canon of Winchester, late Fellow of Queen's College, Oxford. Second Edition, carefully Revised. 4to. cloth. 25*s.*

" *For all earnest students of the Old Testament Scriptures it is a most valuable Manual. Its arrangement is so simple that those who possess only their mother-tongue, if they will take a little pains, may employ it with great profit.*"—NONCONFORMIST.

Yonge (Charlotte M.)—SCRIPTURE READINGS FOR SCHOOLS AND FAMILIES. By CHARLOTTE M. YONGE, Author of "The Heir of Redclyffe." FIRST SERIES. Genesis to Deuteronomy. Globe 8vo. 1*s.* 6*d.* With Comments. Second Edition. 3*s.* 6*d.*

SECOND SERIES. From JOSHUA to SOLOMON. Extra fcap. 8vo. 1*s.* 6*d.* With Comments, 3*s.* 6*d.*

THIRD SERIES. The KINGS and the PROPHETS. Extra fcap. 8vo. 1*s.* 6*d.* With Comments, 3*s.* 6*d.*

Actual need has led the author to endeavour to prepare a reading book convenient for study with children, containing the very words of the Bible, with only a few expedient omissions, and arranged in Lessons of such length as by experience she has found to suit with children's ordinary power of accurate attentive interest. The verse form has been retained, because of its convenience for children reading in class, and as more resembling their Bibles ; but the poetical portions have been given in their lines. When Psalms or portions from the Prophets illustrate or fall in with the narrative they are given in their chronological sequence. The Scripture portion, with a very few notes explanatory of mere words, is bound up apart, to be used by children, while the same is also supplied with a brief comment, the purpose of which is either to assist the teacher in explaining the lesson, or to be used by more advanced young people to whom it may not be possible to give access to the authorities whence it has been taken. Professor Huxley, at a meeting of the London School Board, particularly mentioned the selection made by Miss Yonge as an example of how selections might be made from the Bible for School Reading. See TIMES, *March* 30, 1871.

LONDON: R. CLAY, SONS, AND TAYLOR, PRINTERS.

www.ingramcontent.com/pod-product-compliance
Lightning Source LLC
Chambersburg PA
CBHW021501210326
41599CB00012B/1096